I TOOK THE SKY ROAD

Also from Wildside Press

Hugh B. Cave

Drums of Revolt
Larks Will Sing
The Witching Lands: Tales of the West Indies

Michael Bracken

Bad Girls: One Dozen Dangerous Dames
Deadly Campaign
Tequila Sunrise

Reginald Bretnor

Decisive Warfare: A Study in Military Theory

David Dvorkin

The Cavaradossi Killings
Time for Sherlock Holmes

Joe L. Hensley

Robak's Cross
Robak's Fire
Robak's Firm
A Killing in Gold
The Poison Summer
Song of Corpus Juris
Final Doors
Rivertown Risk
Outcasts

Marvin Kaye

My Brother the Druggist
My Son the Druggist
Bullets for Macbeth
A Lively Game of Death
The Soap Opera Slaughters
The Country Music Murders
The Laurel and Hardy Murders

Ray Faraday Nelson

Dog-Headed Death

I TOOK
THE SKY ROAD

Comdr. Norman M. Miller, USN

as told to Hugh B. Cave

WILDSIDE PRESS

I TOOK THE SKY ROAD

Published by:

Wildside Press
P.O. Box 45
Gillette, NJ 07933-0045
www.wildsidepress.com

First Wildside Press Edition:
June 2001

10 9 8 7 6 5 4 3 2 1

"BUS" MILLER

BY TED STEELE, LIEUTENANT, USNR
AIR COMBAT INTELLIGENCE OFFICER
BOMBING SQUADRON 109

COMMANDER NORMAN MICKEY MILLER likes to fly and has spent more than six thousand hours at the controls of airplanes. He likes the Navy, too. It is his life. But most of all, perhaps, he likes people.

The feeling is mutual. People, all kinds of people, like "Bus" Miller.

A legend began to grow up around him during his combat cruise in the Central Pacific as commanding officer of Bombing Squadron 109. Even to seasoned airmen his personal exploits were breathtaking, and under his leadership his squadron established the best record of destruction against enemy shipping and island bases of any land-based Navy search squadron in the Pacific.

Commander Miller's share of this record was very great. He set the pace for his aggressive pilots with a brilliant succession of single-plane masthead and treetop-height attacks that earned for him, from the Navy itself, the title of "One-Man Task Force."

Commander Miller flew many more hours and many more missions than any other pilot in his squadron. Invariably he took the most dangerous flights. That a legend should have grown up around him is understandable, for he flew unescorted and made his forays on his own initiative after the assigned tasks of the squadron were accomplished. He attacked with impudent daring the most heavily defended enemy targets—including the now infamous island of Iwo Jima. All these attacks were successful.

Commander Miller is not a reckless man, however, and he certainly was not risking his life in a hunt for glory. He did these things because he is a skilled and clever pilot who believed thoroughly in the soundness of his tactics and in his ability to outthink the enemy.

As a commanding officer he was completely informal. He likes to laugh, and so he laughed a lot. His affection for his men was real, as was theirs for him. His leadership was of the personal kind, based on example. He needed no soap-box.

The men of the squadron do not subscribe completely to the Miller "legend," for "Bus" Miller is as real to them as the charts they studied and the planes they flew, and they have come to believe, as he does, that the best way to win a war is to hit the enemy wherever, whenever, and as hard as you can. They admire him as one of the truly great combat aviators of this war. They know him to be an exceptional pilot and officer—a man whose naval career seemed destined for greatness even without his remarkable combat record.

"Bus" Miller is a hero to his own men—who appraise the work of heroes most critically. He happens to be, also, a much loved guy, which in the long run is probably a good deal more important.

FOREWORD

WHEN tales of the almost unbelievable exploits of Liberator Squadron VB-109 began filtering back from the Central Pacific last summer, it became obvious that someone was going to have to do a book about Commander Norman Mickey Miller.

With astonishing efficiency, Commander Miller and his men were removing Japanese shipping from the Pacific, putting enemy islands out of business, destroying enemy planes, and obtaining priceless low-level photographs for our invasion forces. The commander himself was being called "the one-man war on Truk," "the one-man air force," and "the one-man task force." A legend was growing up around him.

As a search and patrol squadron, VB-109 was not required to carry out these dangerous raids on enemy strongholds far from our forward bases. Their official duties were considerably less hazardous. But because they thought it ought to be done, they hammered away at the Jap's defenses day and night, through the Gilberts, the Marshalls, the Marianas, the Kazans, and the Bonins, developing and perfecting an incredibly effective low-level type of attack which has since become an inspiration to naval aviators who followed them.

After nearly eight months of this, VB-109 was sent home for reorganization and reassignment, and more of the details of "Bus" Miller's personal exploits became known. His plane, the famed *Thunder Mug*, had sunk or damaged 66 enemy ships, not counting sampans, barges, and other "small fry." He was the first to take a lone Liberator at masthead height into the Japanese mid-Pacific fortress of Truk, which he did not once but seven times. He waged a one-man war of attrition against Ponape, and another against Puluwat, in both of which the Japs came out second best. His was the first land-based bomber to strike from·newly seized Saipan, in

vii

the Marianas, and the first to attack, at treetop altitude, the inner-fortress islands of Iwo Jima, Chichi Jima, and Haha Jima.

"Bus" Miller had earned the Navy Cross, four or five Distinguished Flying Crosses, five or six Air Medals (his awards are not straightened out even now, and honors for his forays in the Bonins and Kazans are still pending), as well as the Purple Heart and two Letters of Commendation. He wore two American Defense Medals, the American Theater Campaign Medal, and four Asiatic-Pacific Campaign Medals. He had been recommended for the Congressional Medal of Honor, the nation's highest award, for his "unparalleled initiative, aggressiveness and fearlessness, his rare knowledge of enemy tactics, his brilliant and imaginative planning, and the consummate skill, daring and sheer bravery displayed in the execution of his attacks." Yet "Bus" Miller, who has been decorated more times than any other aviator in the Navy, is 36 years old and the father of five children.

When I went to Jacksonville, in October, to talk to Commander Miller about I Took the Sky Road, I did so with misgivings. He hadn't much time, and he had to close up his Jacksonville home, travel across the country to San Diego with Mrs. Miller and the children, and find a home for them there before reporting for duty again. If the story of the commander's career were to be solely a review of the record of VB-109, all well and good; the writer could follow "Bus" Miller around for a few days, pick up the squadron's record at the Navy Department, talk to some of the other men of VB-109, and put together a book. But it had to be more than that. It had to be the whole story of Naval Aviator No. 4050.

For Commander Miller had been flying a long time, and behind the dramatic tale of his wartime accomplishments lay an important and equally stirring story of the years of training that go into a career in naval aviation. "Bus" Miller had been graduated from the United States Naval Academy in 1931, from the ranks. He had flown from carriers when the Navy had but four of them. He had piloted catapult planes from warships, instructed at Pensacola and Jacksonville Naval Air Schools. He had flown PBYs with the first Neutrality Patrol, when the Atlantic was swarming with submarines. He had been commander of the experimental flying boat, XPBS-1, which blazed the trans-Pacific air route to Australia and Java during the desperate days immediately following the Japanese attack on the Hawaiian Islands.

In short, his achievements as Commanding Officer of VB-109 were the climax, but not by any means the substance, of a long, exciting, adventurous career in naval aviation—a career in which American young men should be vitally interested.

So in October, afraid that in the limited time available it would be almost impossible to get to know "Bus" Miller well enough to do a book about him, I arrived in Jacksonville. I needn't have worried. You get to know him very easily, and you quickly understand why his men think there is no one else on earth quite like him. Behind the Miller legend is a tall, soft-spoken, friendly man with large hands and feet, a big nose, an amazing capacity for work, and an utter lack of anything that might pass for conceit. He insists that the Miller "legend" is a lot of nonsense, and that he and the rest of the pilots and crews of VB-109 only did what they were trained to do.

Describing the Miller face is a difficult job. Its expressions are so varied that in trying to select a photograph of him for a pictorial record of the squadron, his own officers had to use not one picture but six, and all are different. All, however, have one thing in common: they show the commander's easy good humor, his warmth of personality, and his deep natural dignity. Under the six pictures in the squadron record, his men placed the following caption:

"Commander Norman Mickey Miller, USN, Captain, Old Man, Bus, Skipper, the One-Man Task Force, skilful aviator, center of myth and legend, nerveless, barber-shop tenor, sunbather, nighthawk, party spirit, hater of altitude, aggressive fighting man and good companion, inspirer within the squadron of alternate anxiety and awe—and great guy. No foolin'!" That's tribute enough, and from men who know.

Here, then, is the "Bus" Miller story. I have not "written" it, in the sense that many such stories are written from scraps of official information and bits of conversation. I have simply set down and arranged the scores of stories he told me—with the same quiet dignity, I hope, in which they were told. Wherever I have thought it important, for the sake of the record, to comment on the commander's remarks or to include comments by others, I have carefully followed the policy of identifying these "asides" for what they are.

HUGH B. CAVE

CONTENTS

CONTENTS

THE most remarkable thing about my father, W. F. Miller, was the alert, unselfish interest he took in his eleven children. He was a busy, energetic man, a builder of mills and mill-towns, dams, houses and large commercial structures. His work kept him on the jump at all hours. When he was out, we seldom knew where to reach him. When at home he was usually deep in paper work or at ease with a good cigar.

As a boy of thirteen with only a few years' schooling, Papa had a job picking up chips at twenty-five cents a week. By plain hard work through the years he went on to be vice president of a large, established construction company. To all of us he was an inspiration.

He was more than that, really, for with all his activities he devoted more time to his brood than most fathers give to a child or two. He knew how we were progressing in school, how and where we spent our spare time. He knew our associates, our hobbies, what we were reading, and more often than not, what we were thinking.

In my case he knew, long before I was certain myself, that I wanted to go to Annapolis.

Papa thought it was a good idea, and so did Jack, the brother next above me in age. No one in the family had ever served in the Navy and we had no naval connections whatever, but that was of small importance. "If it's Annapolis you want," Papa said, "let's see what we can do about it."

We lived in Winston-Salem, North Carolina. (When I was born the hyphen had not been inserted and the town consisted of two separate communities, of which we lived in the southern or Salem part.) We were members of the old Home Moravian Church, and Papa knew nearly everyone in town. With simple confidence he set out to obtain for me an appointment.

It troubled him, I think, when he discovered that appointments to the Academy were not obtainable at the corner drug store. He was guileless and direct by nature; complexities annoyed him. He used to say to those who tried to disillusion him, "Bus has a good school record, even though he never did study very hard. He's earned good enough marks. Why shouldn't he go to Annapolis if he wants to?"

As a matter of fact my marks were pretty good—at or near the top of the list at the local high school—and I hadn't had to burn much midnight oil to get them. School had never caused me concern. Most of my homework was done as a sort of side-line, a necessary but not unpleasant evil, while I sat in my room at the radio.

You remember the radios we had in those days, with headphones, enormous panels, and more dials than a locomotive has wheels. Mine occupied two or three tables, the bureau, and usually most of the bed. On the wall was a huge map of the United States, and night after night during my senior year at high school I sat there with the phones on, logging stations. Each new station was a stupendous event. The map was weirdly and wonderfully patterned with colored threads stretched between Winston-Salem and the radio stations I had "discovered." "Hey!" I would shout, racing downstairs in the morning. "I got KFI last night!" "Well, now, that's fine," my mother would say. "But you must quit sitting up so late over that radio or you'll ruin your health. And what about your studies?"

I must have done my Latin, history, and other studies between stations, or else the teachers at Winston-Salem High were more fond of me than they ever let on. Possibly they just didn't relish the thought of another year of my trombone playing in the band. At any rate my marks were good, and Papa thought with commendable parental pride that they were good enough for Annapolis. The trouble lay in the fact that Winston-Salem, with its population of about 65,000, was apparently devoid of people with Annapolis connections.

Eventually Papa went to see Lawyer Sams, a state congressman who lived in my home town. Papa had done a lot of building for him. In return, Lawyer Sams had said, "Will, if ever you want to get one of your boys into Annapolis or West Point, let me know and I'll do what I can to help." And Mr. Sams did just that. He went right to work on an appointment for me.

Actually I was not too concerned. With school over I had gone to work for my father, who was then building an apartment house. I had money in my pockets and spent most of my spare time riding hell-for-leather around town in a jazzed-up Model T Ford which Papa had bought second-hand and I had converted into a rampant chariot that frightened chickens and old ladies out of their wits. With wings, that explosive Model T would have been the first low-level bomber. And then, too, I had met a girl.

Thelma Davis was a year behind me in high school and sang in the glee club. I played in the band. We met for the first time at a music festival at Greensboro, in which we were both participants. There was dancing afterward, and when I saw her dancing with other boys I promptly abandoned my own date and pestered a friend to introduce me. "I want to dance with that girl," I told him. "I've got to know her."

Until then I hadn't paid a whole lot of attention to girls. My spare time had consisted of tooting the trombone, logging radio stations, helping Papa and wondering if I'd ever get to Annapolis. But at the Greensboro music festival I eventually wangled a dance with Thelma Davis, and when the evening was over we ditched our dates and went home together. Before long we were "going together" in earnest, and I had lost a lot of interest in trombone playing and the radio.

One dark evening we were out riding in the Model T with a pint of liquor on the seat between us. This was during Prohibition, and while it was considered smart for high school kids to have a bottle on the hip, the pint was actually not ours; I had agreed to deliver it for a friend. Some distance from town, as we rolled briskly down a steep, curving hill toward a narrow bridge, the jalopy's lights went out. A car was speeding toward the bridge from the opposite direction.

I leaned on the horn, but that was defunct too. I stepped on the brakes and managed to coax the Model T to a stop, but it was impossible to pull off the road; there just wasn't room. Jumping out, I frantically waved my arms and yelled.

The oncoming car halted about ten feet from me, and out of it stepped a tall, stooped figure, brandishing what appeared to be an old-fashioned horse pistol. "All right!" he said. "Stand right where you are! Come a step closer and I'll shoot!"

Too scared to take a step in any direction at all, I simply stood there with my mouth open. There had been rough stuff on the roads lately—bootlegging, hijacking, that sort of thing—and when I thought of the bottle of liquor in the Model T, beside Thelma, my knees knocked. Still brandishing the pistol, the fellow warily walked toward me. He was a deputy sheriff.

"What's the trouble?" he demanded suspiciously.

I explained to him that the ignition, lights, horn, everything, had gone bad.

"Will she run?" he said. "If she'll run, you can at least get her off the road."

In trepidation I followed him to the Model T, knowing that if he found the bottle on the seat we'd be in real trouble. Anti-liquor sentiment was decidedly dark brown in those days, and high school kids with whiskey in their possession were headed for perdition whether they meant to drink the stuff or not. But I needn't have held my breath so hard. When we reached the car, the bottle had disappeared. Only Thelma was there, shaking her pretty head and saying with a sigh, "It won't run, I just know it won't."

It wouldn't, either. The sheriff tried and so did I, but the old Model T had evidently made up its mind to lie down and die right there on the highway. While we were pushing it off the bridge another car came by. The driver promised to send out a tow from town.

Awaiting the tow truck, Thelma and I sat on a guardrail beside the road, and the sheriff perched there with us. He was not suspicious any more, but apparently had nothing better to do and enjoyed our company. So there we sat, the three of us, solemnly talking about bootlegging, hijacking and the evils of alcohol. It was a learned discussion, I'm sure, but at least one of us did not appreciate it. By now, having discovered at last where the bottle was hidden, I was running a temperature.

Eventually the truck came, the tow-rope was made fast to Lizzie, and we bade our deputy-sheriff friend farewell. As we climbed in and took off, he waved. Obviously he considered us nice respectable people—which, in fact, we were. (But in those days it was more of a crime to transport liquor than to drink it!) Down the dark North Carolina road we went, our sighs of relief smothered by the rumble of the tow-truck. And then, modestly turning her back to me, Thelma thrust a hand down the neck of her dress and extracted the pint.

"Honey," she said, "that was too close. You'd better go to Annapolis before we both wind up in the clink."

Oddly enough, I didn't wind up in the clink until I got to Annapolis. But that's another story.

As I've said, I worked for Papa that summer after graduation from high school. Like most of the people who worked for him, I thought him wonderfully fair. He never asked a man to do a chore that he would not do himself, and when he criticised, which was seldom, he didn't scold but came straight to the point. One evening, for instance, I neglected to turn some wheelbarrows over. During the night it rained and when Papa arrived on the job in the morning he found the Georgia buggies, as we called them, full of water. He called me, turned one of them over and said, "Norman, rusty barrows fall apart pretty fast." Then without another word he went about his business, leaving me to empty the rest of them. (Papa always called me by my first name. Everyone else called me "Bus" except my mother. To her I was "Brother.")

One day at the apartment house we were building, the foreman sent for me. "Run down to the hardware store," he said, "and get some doorknobs I ordered." Always eager for an excuse to go for a ride, I rattled downtown in the Model T and hunted a place to park, locating one finally in front of the post office. When I jumped out of the car I stumbled against one of those metal sidewalk stands on which the Navy displays posters. I stood there looking at it. Presently an Army man, a sergeant, approached from the post office steps and spoke to me. (If this sounds like a cooked-up story, I'm sorry. It really happened that way.)

"You thinking of joining the Navy, bub?" he asked.

I replied vaguely that I didn't know. Until that moment it had not occurred to me that I might get into the Navy by enlisting; the Navy had meant only Annapolis. The sergeant—his job was recruiting—was willing to answer questions, however, and knew more about the Navy than many a Navy man I've met since. No recruiting was done in Winston-Salem, he informed me. I would have to go to Raleigh. I told him I had to get some doorknobs first and left him, but on the way back to the apartment house I pondered what he had said.

That evening after band practise I didn't feel like going straight home. It was a warm June night, still light at eight-thirty. With my friend and fellow band-player, Archie Spaugh, I checked around in the Model T to look up the girls we knew. We sang the songs and marches we had just rehearsed, I bellowing trombone noises through my nose while Archie clapped his hands and beat out the um-pahs. In a festival mood we waved and tooted the horn at people we passed.

After a time this aimless cruising began to pall and I turned the jalopy into a side street. "Archie, have you ever thought about joining the Navy?" I asked.

"Why should I join the Navy?" he said. "Besides, I thought you were going to Davidson." (My brother Jack was a student at Davidson College and I had enrolled there, for my chances of obtaining an Annapolis appointment were slight, and I knew it.)

"I don't want to go to Davidson," I said. "I want to join the Navy."

"You mean just as a sailor?"

"Well, I can't wait forever to get into the Academy. Besides, you can go to Annapolis from the ranks if you pass the exams."

"I think you're crazy," Archie declared. "What if you don't pass the exams?"

We argued like that for an hour or more because it was something to do and we were tired of riding around. Of course we got nowhere because Archie had no intention of joining the Navy and I was not sure what I wanted. I knew only that waiting around day after day for the appointment was making me restless and unhappy. Eventually we talked ourselves out and I went on home. It was about midnight when I went up the stairs to my room.

The family was asleep and the house was strangely quiet. Perhaps it was the stillness that finally influenced my thinking. All evening while arguing with Archie I had been more than half sincere, and now it occurred to me that everything I had said made sense. I *could* join the Navy and try for Annapolis later. If I failed to make it, at least I would be in the Navy and that, after all, was what I wanted.

In pajamas I went down the hall to my parents' room and put on the light. Papa, never a deep sleeper, awoke instantly and asked what I wanted. Standing by the end of the old-fashioned bed, I said as simply as I knew how, "Papa, I've been thinking it over. I want to join the Navy." And I repeated what the recruiting sergeant had told me.

My mother was awake then, too, but she let Papa do the talking. After frowning at me for a moment he said, "What about your appointment?"

"It may never come through. I just want to be in the Navy."

"Are you sure?"

"Yes, sir, I'm sure."

"Well," he said, "set your alarm clock and we'll go to Raleigh first thing in the morning."

Lieutenant Steele, in charge of the Navy's recruiting office at Raleigh, was an old Navy man, short and muscular and wonderfully understanding. We walked in on him about eleven o'clock after the long ride over the road from Winston-Salem. He shook hands and said, "Sit down. What can I do for you?" When he saw me stealing a glance at the pictures on the wall—they were mostly pictures of submarines—he added proudly, "I served aboard some of those subs, son. I was in the submarine service a long time."

Lieutenant Steele's job was to talk to kids like me, who thought they wanted to join the Navy but were bewildered by the enlistment routine and hesitant to make the plunge. He knew his job well. Before asking a single question he told us something about himself, the high spots of his career, and made us feel that we had known him all our lives. Then he looked at my report cards and letters of reference and listened sympathetically to Papa's tale of the appointment and my desire to enter the Naval Academy, which was known to us landlubbers then as simply "Annapolis."

"How old are you, Norman?" he asked me.

"Eighteen."

When were you born?"

"February first, 1908."

A look of concern crossed his face and he began to do some mental arithmetic. "You haven't much time if you want to make Annapolis," he finally said. "To get in from the ranks you have to be under twenty and must have served a full year in the Navy by July first of the year you enter. Today's the thirtieth of June and you're eighteen. If you haven't served your year by the first of next July, you'll be too old. See what I mean?"

Papa did. For a moment he frowned at the calendar on the lieutenant's desk. Then, turning to me, he said, "Son, you'll have to make up your mind right quick. If you're not in the Navy by midnight tonight, you'll never see the inside of Annapolis."

I had it straight after a while, and suddenly was panicky. Until that moment there had been a way out. I could have thanked the lieutenant for giving so much of his time and told him I would think it over. I could have gone home with Papa to make up my mind. Now there wasn't time for that.

The lieutenant, sympathetic, advised us to take our time about reaching a decision, and left us. Papa said, "It's up to you, son. The appointment is still pending, you know. It may come through if you wait."

"But it may not," I said miserably. "Then I *will* be sunk."

"Well, it's up to you."

"I'd better sign up now," I decided.

The actual signing wasn't quite that simple. First we had to secure affidavits from three people testifying to my good character. Friends in Raleigh were willing, but it was nearly three o'clock when we arrived back at the recruiting station. I was sworn into the United States Navy by Lieutenant Steele at 3 P.M., June 30, 1926, and though I had asked for it and wanted it—and have never since regretted it—I was almost in tears when I lowered my hand.

One part of that signing-up procedure was worth a smile, if I had been in a smiling mood. It occurred while I was fretfully waiting for the physical examination. Into the room came a squat, grizzled old boatswain's mate with hash marks (enlistment stripes) almost to his shoulders. He sat down and hunched himself toward me. "Son," he said, in a voice like a bullfrog's, "let me give you some words of advice. When you get to the Academy, study all you can about sailing ships and sailing seamanship. That's mighty important. Trouble with the Navy today, we've been neglecting sails and depending on steam too long. Now me, I was on sailing ships and I know what I'm talking about. You do what I tell you."

I promised him solemnly that I would. At that point I would have promised anyone anything!

The lieutenant knew how I felt. Before the ink was dry on my enlistment papers he was quietly telling me what to take to boot camp, what to expect when I got there, how to conduct myself, and all the rest of the "dope" on how to get along in the Navy. He was proud of the Navy, proud of his years of service, and proud, too, of the job he was doing at the moment. When Papa told him that my mother was ill and this would be a shock to her, he leaned back in his chair and said, "Well, now, according to regulations you should leave for Norfolk this afternoon. But I think we ought to consider your mother's feelings. Suppose you go home, break the news to her, and report here tomorrow morning." When Papa shook the lieutenant's hand he really meant it—because this was a shock to him, too, despite his determination not to show it.

After we had told my mother that night, I called on Thelma and took her riding. But if I had any romantic anticipation of a fond farewell, I was mistaken. The longer we rode, the less we talked. The more we talked, the faster fled romance. Thelma had admired my ambition to go to the Academy, but not so alluring was the prospect of being "engaged"—even high-school "engaged"—to an ordinary seaman. It was a shock to her. My joining the Navy had shocked just about everyone.

I hadn't been fair to her, she said. I hadn't considered her feelings. I didn't love her and she didn't care whether I did or not. If I wanted to join the Navy I could go ahead and *join*. (As if I had any choice!)

With the same kind of eighteen-year-old logic I vehemently declared that she ought to be proud of me in any kind of Navy uniform. Furthermore, I'd get into Annapolis if it killed me, just to show her.

That was our fond farewell. Girls, I had decided, were just plain ornery. No more girls for Bus Miller. If ever the moon tempted me again I would tilt my head back, a lone wolf, and howl at it.

2   *SABERS AND PRAYERS*

MOST of what is written about boot training today is concerned with the training of war-time inductees. Forgive me if I seem to sound like a nostalgic graybeard lamenting the "good old days," but the system at Norfolk in 1926 was considerably more rugged. I can show scars to prove it! Naval trainees then were all volunteers, having of their own free will put their necks out. The usual retort to any complaint of rough treatment was, "Listen, mate. No one pushed you into this man's Navy, so pipe down!" We did a lot of piping down. Sounding off was apt to be dangerous.

Norfolk at that time was really run by the chiefs who were stationed there, most of whom were grizzled old-timers with their own notions of what boot training was for. The man I most remember was Chief Boatswains Mate Hall, a grand, human guy when we discovered the chinks in his armor, but so blistering at first that he frightened most of us half out of our senses. His job was to transform the "men" of Company 16, my outfit, into sailors. He hadn't much to work with. Most of us were physically limp and mentally rigid. Some were mere infants of fifteen. One secretly admitted he was only fourteen.

The chief did his best in the way that seemed best to him. When our drilling didn't suit him, he would march us across the training station to an area called Unit X, used during the First World War as a training ground and later abandoned. It was a desolate spot, remote enough from the station proper to suit the chief's desire for privacy. In the midst of a desert of ankle-deep sand stood an ancient barracks building, left to rot on its foundations, and behind it, where he could have us quite to himself with no fear of snoopers, the chief made us do double time under arms in a tight rectangle.

For hours on end we marched in the hot sun of July and August while the chief stood in the center like a ring-master and whanged the laggards with his saber. Some of us, I'll wager, still wear the welts of that weapon. All of us remember Unit X and the miseries we endured there.

Yet the chief won Company 16's profound respect with this sort of treatment. He shaped us into Navy men in a way that stuck. Now and then I still run across some of Company 16's veterans, scattered today all over the Navy, and to the question "Remember Chief Hall at Norfolk?" the answer is certain to be, with a grin, "That old son-of-a! He sure knew his stuff!"

Navy trainees bunked in hammocks then, and the hammocks at Norfolk, slung between jack-stays in the barracks, were something to remember. They had to be taut. None of your back-yard slackness, with a comfortable canvas cradle slung for repose of the soul! If our hammocks were not as tight as fiddle strings the chiefs passed out penalties with abandon.

Of course we were forever falling out of the things, and it was a rare night when some poor devil didn't hit the deck with a tremendous clatter. Two of the boys broke their arms in nocturnal nosedives. Another, a red-haired lad from Alabama, struck his head with such force that he blacked both eyes and flattened the whole front of his face. And I recall great turmoil one night when one of the boys, evidently epileptic, had a fit. He rolled out of his hammock, hit the deck with a din to wake the entire barracks, and swallowed his tongue. The rest of us, wondering what to do, stood helplessly by and watched him get blue in the face, until someone pulled the poor fellow's tongue back out to keep him from strangling. That night he was taken to the Naval Hospital, and that was the last we heard of him.

Cleanliness was the by-word there. We had to do our own washing, of course, and there were numerous bag inspections when everything had to be laid out just so, or else. If a boy became slack, we went to work on him, sometimes on our own initiative, sometimes in response to strong hints from the chief. The chief diplomatically turned his head while we led the culprit out to the washrack, stripped him down and scrubbed him unmercifully with ki-yis (Navy scrubbing brushes) and sand. The cure was always effective.

I had been at Norfolk about two weeks when a letter came from Papa. I had received an appointment to the Academy! To make it secure I had only to take certain substantiating examinations—nothing like the stiff regular exams that would confront me if I tried as an enlisted man to enter the Naval Academy Preparatory Class for admission to Annapolis the hard way. There were the usual complications, however. I would have to get out of boot camp and the Navy first, and this might take some doing. It could be done, though, by special order discharge. Papa made it plain that the decision was up to me.

I pondered all the angles before answering the letter. At Norfolk, despite the man-killing hammocks and Chief Hall's saber, I was not unhappy. In two weeks I had made a number of friends. I had gained weight and been toughened physically and mentally. I liked the life. Furthermore the chiefs at Norfolk had already convinced me that I knew pathetically little about the Navy, and I thought it would help me to know more before going to the Academy.

One of the boys finally made up my mind for me. "If you can't pass the exams for Prep Class," he said, "you probably wouldn't pass them on the outside and wouldn't belong in Annapolis anyhow."

"Papa," I wrote, "please thank the Congressman and tell him I have decided to stay here. I'm in the Navy and that's what counts. There's no telling what might happen if I cut loose now."

Obviously pleased, he wrote back: "Whatever you think best, son." So I stayed at Norfolk, finished boot training, and with two other boys in Company 16, Ed Bingham of Norfolk and Tom Henry of New Bern, N. C., passed the entrance exams for Prep Class. Thirteen of us tried for it; three of us made it. My troubles, I felt, were about over.

Life in the Naval Academy Prep Class (NAPC) at Hampton Roads was only a little less rugged than at boot camp. Our quarters were single-story wooden barracks which had been constructed inside and out by men who passionately disbelieved in frills, and though we had cots now instead of the boot-camp hammocks, we soon learned that Virginia in winter can be cold, especially when you entertain it indoors!

The principal difference was the type of work we did. At boot camp we had labored hard and long under Chief Hall to learn to be sailors. At the NAPC, striving to be students, we battled mountains of books and another tough but good-hearted old chief boatswains mate, "Pop" Ellis.

One day in November—the thirteenth, I think it was—some of us went over to watch the Schneider Cup Races. We had no secret urge to be aviators, but the Schneider Cup affair was an important aviation event attended by spectators from all over the country. Leading fliers of the U.S.A. and Italy took part, and the competition was internationally intense. England, that year, was not represented.

The day was blustery and raw, November weather at its worst, with a squally wind turning up off Chesapeake and Willoughby Bays. With our hands stuffed into our pea-coats and our feet in constant motion, we shivered for hours. The Navy of course was well represented in the crowd, and before long, everywhere you looked, the Navy's nose was running.

Italy took the prize that day, and an Italian, Major Bernardo, broke the world's speed record in a little red low-winged monoplane on floats, doing a fraction over 246 miles an hour. That was *fast!* When it was announced, the crowd gave out a great gasp and then cheered. As for me, I was more interested in the red monoplane than in the man who had flown it. For the first time in my life there was awakened within me a curiosity about "wind machines"—those man-made things that flew in the air. "Come on," I urged the boys. "Let's have a look at that plane."

We got fairly close to it by some expert line-plunging through the crowd. I thought it a beautiful little thing, sitting there on the field. I wanted to touch it, sit in it—but not to fly in it. The idea of being 'way up there with only an engine and a wing to keep me there was more than I cared to contemplate. But on the ground it was beautiful, and with our noses dripping and our faces turning blue we stood there in the open, stung by the wind, admiring it.

That night I awoke in a weird sweat and knew I was ill. It was all I could do to get out of my bunk and call the chief. He took one sleepy look at me, said, "Miller, you aren't kidding," and hustled me to sick bay for a check-up. Half an hour later I was aboard the hospital boat, being transported up-river to the Portsmouth Naval Hospital, with scarlet fever. By morning there were four other boys from the NAPC in the same ward.

We really had our troubles then. A new type of serum was in use at the time—at least it was being used experimentally—and the good doctors at Portsmouth jabbed us full of it. It made all of us terribly sick and raised enormous welts on us. Perhaps it was supposed to do that—I don't know—but for four days we lay in profound misery, with melons sprouting out of us, expecting confidently to die. Then the lumps subsided and we were suddenly quite well again.

We were still under quarantine, however, and though permitted to sit up and talk to the nurses, we were not allowed to read.

Reading in such weakened condition, we were told, might permanently injure our eyes. That was bad. Getting through NAPC was difficult enough without being denied textbooks. We counted the lost days, six weeks of them, and each seemed a mountain in the way of passing our exams for the Academy.

"This makes it certain we won't get to the Academy," we said miserably. Really believing it, we resolved that when released from the hospital we would go out in a group, high, wide and handsome, to get "stewed, lewd and tattooed"—forgive me if I alter that quotation a bit, but any Navy man will know what is meant —and thereby sink our sorrows. The resolution wavered somewhat when on Thanksgiving Day we were treated to a full-course dinner and told that our stay in the hospital would soon be over. A few days later, however, I learned the real meaning of sorrow.

My mother had been at a hospital in Philadelphia, undergoing an operation. She had not been well for some time, and the stay at the hospital, it was hoped, would put her on her feet again. One morning Miss Templeton, the nurse in our ward, came and stood beside my bed and told me my mother had died.

A small-town boy with family ties as tight as mine should be older than eighteen, I think, when faced with such a loss. For days I was completely sunk. Had I been able to go home to my people it might have been different; as it was, I moped about the hospital refusing to be consoled. Yet I think the death of my mother did much to hasten my "coming of age," for after one fling which failed to come off as planned, I settled down to business and studied desperately to make up for lost time. The fling came in December, when I still believed that my world had come to an end.

I had been telling myself that the way to forget was to do what we five in the ward at the hospital had planned to do—go out on a tremendous binge. As a matter of fact I had never done any binging. At the NAPC we had little time to ourselves and rarely got to town. When we did go it was usually to a show or a pool parlor. Prohibition of course was still in effect; the only available liquor seemed to be a lethal kind of white corn which made me sick the first time I tried it, and which I left strictly alone thereafter. Once or twice I had gone along with the boys when they went adventuring, but I was genuinely afraid of the corn and stuck to coke or milk. Being a small town boy at heart, I was afraid of other things as well, but you could always kill time waiting for your buddies by playing the phonograph or conversing with the "madame."

This afternoon, though, I had resolved to bury my inhibitions and "hit the jug" if it scuttled me. My eighteen-year-old philosophy insisted that only through a complete mental and moral change could I forget my troubles. Moreover it was New Year's Eve, the accepted time for such things. So with my pals Tom Henry and Dick Helm—the latter from Bristol, Tennessee—I headed for Norfolk and the bright lights.

It didn't work. The corn liquor was no more genteel on New Year's Eve than when I had tried it before, and I was soon gulping cokes again to keep from being sick. As for the second resolution, we were all rank amateurs and our courage fled at the threshold. Disgusted with ourselves, Dick and I swore to each other that at least we would get tattooed, and so at ten o'clock we headed for Main Street and Coleman's. Tom Henry said "To hell with it" and steered a course for the Navy YMCA.

Every sailor who has ever set foot in Norfolk knows Coleman's. Still we entered with trepidation. Dick, who had more courage than I, was operated on first. Handing the artist a letter he had received that day from his girl, "Liz," he said, "That's her signature; copy it on my arm." The result was gaudy if not artistic.

I had no letter, but the choice of designs on Coleman's wall was practically limitless. You could have mermaids or battleships, bleeding hearts, undraped females, numerous tributes to "Mother," or any of a hundred and one other luscious items. With fading fortitude I pointed to a "USN" superimposed on an anchor, asked the tattooer to add a " '26" at the bottom, extended my arm and squeezed my eyes shut. An hour later, long before the tattoo had healed and the arm had stopped smarting, the date was '27. Next day we regretted our visit to Coleman's, but as Dick sheepishly pointed out, at least we were only tattooed!

In April our sojourn at the Academy Prep Class ended and we struggled with examinations. I remember six of them—American and ancient history, English, algebra, physics and geometry. After each one I regretted more than ever that I had not snatched the chance to enter Annapolis the safe way, by appointment.

Still the prospect of failure did not really upset me. The Navy had been good to me; I liked what I had seen of it. Win or lose, what did it matter? Most of the boys expected to flunk out anyway, and I would have plenty of company. (Today, looking back, I find it as difficult to understand that philosophy as to anticipate the antics of Jackie, my ten-year-old son. Recalling that early "hell-with-it" attitude, I break out in a rash of goose-pimples!)

The awaited verdict was not handed us immediately. At the conclusion of the exam period we were finished at Hampton Roads and told to go home, the decision to be telegraphed to us when the papers had been checked. If we had made the Academy there would be an extension of leave; if not, orders to report elsewhere would be forthcoming. I was so certain of failure that I left my sea-bag at Ed Bingham's house, and my trombone in a Main Street pawn shop for ten dollars, fully expecting to pick them up when ordered back to Norfolk.

At home, possibly to build up a defense against the receipt of bad news, I insisted to Papa and the rest that the Academy didn't matter—I was in the Navy and proud of it, and who cared about being an officer, anyway? This went on for days. Perhaps they believed me, perhaps not. Papa, who probably cared most, never let on. I rode around town in the Model T, went to the movies, called on high school pals and generally killed time, waiting. Thelma Davis had moved from Winston-Salem since my enlistment and was not there to be dated.

"Annapolis?" I said to people who asked questions. "Shucks, they can get along without *me*."

My change of attitude was sudden and dramatic. Once again I

believe the silence of an empty house was at least partly responsible. (With seven brothers and sisters living, I was not exactly
used to a silent house!) Everyone had gone out this evening, and
with nothing to do and no one to talk to, I was moping about
from room to room when the telephone rang upstairs. All day I
had expected to receive the verdict from NAPC. This was the
call—I knew it. No one has ever called me psychic, but I was
certain.

"Telegram," the operator said, "for Norman Miller." And suddenly I lost the protective veneer of indifference. With heart and
soul I wanted the news to be *good*. The kidding was over.

I come from religious people—the family's roots are fixed far
back in the Moravian Church which, as you may know, is one of
the very oldest of Protestant churches and stems from the disciples
of John Huss. From early childhood I had been taught to pray
in moments of crisis, and while there have been times when the
praying was done silently, the habit is still firmly fixed. That
evening, alone in the house, I took the telephone call on my knees,
while the operator matter-of-factly read: "Report to *Reina Mercedes*, Station Ship at Annapolis."

I was in. Sixty days' leave and then the United States Naval
Academy!

3   *ALMOST TOO LONG IN THE BRIG*

DURING leave I went to Washington with my brother Jack and
some friends. It was June, 1927. A fellow named Charles Augustus
Lindbergh had just flown non-stop from New York to Paris in a
plane he called "The Spirit of St. Louis." He arrived in Washington
while we were there, and the city enthusiastically turned out to
welcome him.

For a man who was later to pilot carrier and catapult planes,
four-engined flying boats and Liberators, I was peculiarly unim-
pressed. Airplanes and aviators simply did not interest me, and I

found the fireworks display more exciting than the Lindbergh festivities. The fireworks were really something to see. I watched them from the Lincoln Memorial steps while Jack and the others struggled through the crowds on Pennsylvania Avenue to catch a glimpse of the returning hero.

(I still feel that way about heroes. The Navy, for instance, is full of them. Most of them never wanted to be heroes. And for every one you hear about there are thousands who just don't happen to be around when the correspondents are taking down names.)

Probably my indifference to flying stemmed from boot training at Norfolk, where the noise of the planes had interfered with our studies and caused us to gripe at every naval aviator we met. We had watched the trainees practising carrier landings, too, and there were some nasty crack-ups on the simulated landing deck at the field by our school building. Some of them I remembered all too well. Probably I was scared. At Norfolk, when the boys had hitched air rides and asked me to go along, I had always skinned out of it by recalling an important letter that ought to be answered or work that had to be done. So if this fellow Lindbergh wanted to risk his neck by flying across the Atlantic in a land plane, that was his baby and he was welcome to it! In the Navy I intended to keep both feet firmly on a ship's deck.

On June 29, 1927, one day short of a year from my enlistment, I reported to the Annapolis station ship. The next day, after being an enlisted man for a year to the hour, I was honorably discharged from the Navy on the *Reina Mercedes*, and then marched to Bancroft Hall where I was sworn in as a Midshipman, Fourth Class. The station ship, incidentally, was then commanded by a captain named William F. Halsey, Jr., whom the world and especially the Japanese navy now knows rather well.

I won't bore you with tales of Academy life. There are dozens of books in which that sort of thing has been done better than I could do it. I won my share of demerits, made many friends, and in 1929 had my first ride in the sky.

The plane was a float biplane, a Martin T4M, forerunner of the modern torpedo plane. Lieutenant Ramsey, the pilot, is now an admiral. We flew leisurely over beautiful Chesapeake Bay for an hour or so and leisurely flew back again. I wasn't scared. I wasn't impressed much, either.

Later I was taken up in a twin-engined seaplane known as an F5L, forerunner of today's PBY. This jaunt lasted a little longer. Of the five midshipmen who went along as passengers, three of us—Reginald Rudolph McCracken, "Swede" Larson and I—in devious ways, eventually became naval aviators. Another man, Jerry South, became one of the country's best navigators and did the navigating for the *Bowditch*, when that ship performed some hazardous and important chores in surveying for our new Atlantic bases.

That same year, 1929, I met Thelma Davis again. While at home on leave for a few days, I was told that she had returned to Winston-Salem on a visit. Not at all sure of the reception I'd get, I telephoned. It was a most encouraging reception. Before the evening was over we were "going together" again, this time in earnest, with substantial plans for the future. After seeing her back to the apartment in which she was staying, I drove home on clouds. Oddly enough we later moved into that apartment—but then there were four of us.

Even if I wished to, I could not make the story of my stay at Annapolis very exciting. It was mostly hard work. Football ambitions were partly snuffed out by the recurrence of an injury suffered in high school days, though I did squeeze into some important games. I also made the wrestling team. Other Academy memories are somewhat confused. But one thing I do remember, and vividly, is First Class June Week. Most of it I spent in the brig!

As you probably know, June Week is the midshipmen's gala wind-up. Examinations are over. Navy has licked Army (of course!) in baseball, lacrosse, track and other spring sports. The annual midshipmen's summer cruise is in the offing—maybe it'll be to Europe! Hundreds of the best-looking girls in America are on the scene for the June week dances, and proud parents are arriving for graduation. Every midshipman with a propensity for collecting demerits is on his noblest behavior lest he acquire one too many and find himself out on a limb.

With June Week coming up my demerits totaled 146, far too many for ease of mind. Where and how I had amassed so many I don't recall—and it's probably just as well—but four more might mean dismissal, and there were still five long days between me and my commission. My father and step-mother were due in to attend the graduation.

With some misgivings I appealed to my roommate, "Easy" Stewart. (Walter Jones Stewart, a big, good-looking boy from New Jersey, later saw war-time action while serving in destroyers.)

"Easy," I said, "this is one time old Bus Miller will have to play it cagey. No more demerits for this chicken. If I even so much as step on the grass, I want you to slap me in a strait-jacket."

Easy knew more about demerits than even I did. (It must have been the room we shared!) He had already served time on "the Ship," in the Black Hole of Calcutta, and been branded with the "Black N." But he was sympathetic. "I'll keep an eye on you," he promised. "Not that it will do much good."

"My career in the Navy has caused Papa headaches enough," I said. "I'll be good until graduation if I have to hire a guardian."

"You'd pervert the guardian," said Easy.

All that day I walked on tiptoe, on the alert against demerits. Not a single Academy rule, so help me, did I infract. The following evening Easy had a date in town—so he said—and at dinner I was "in charge of table." Certain that he had done it with a purpose, I could have strangled him for putting me in such a spot, for in June Week the boys naturally let down a bit after the tension of exams and there was likely to be some horsing around. With all its stern demeanor the Academy is still human.

Sure enough it happened. Midway through dinner, with twenty-two men at the table, my charges discovered that by squeezing the bread into soggy little balls and flipping them at one another, they could have fun. It was up to me, of course, to maintain order or suffer the consequences. Uneasily I looked around. The duty officer's table was decently distant and no one there appeared to be aware of the antics. "Oh, well," I thought, "who am I to be a stuffed shirt?" The bread-balls went on flying.

Dinner over we rose as usual, were dismissed and walked out. I'd been lucky, I thought. No one—ha ha!—had noticed the horse-play. But before I reached my room the word was passed for Midshipman Miller to report to the main office. Old Miller was in the soup again.

It went something like this:

"Midshipman Miller, were you in charge of table such-and-such at evening meal?"

"Yes, sir."

"Hmm. I inspected that table and discovered bread on the deck. Did you permit the midshipmen to throw bread?"

"Er—yes, sir."

Brief pause for scrutiny of Midshipman Miller's demerit record, followed by a small sigh. Business of adding 25 more demerits to the list and marking down the new total—171. Then, sadly: "You will report to the ship at once, Midshipman Miller." Meaning, of course, the *Reina Mercedes*, otherwise known as the Black Hole, the prison ship, and a hell of a place to be in June Week.

You can smile at such things after the years have mellowed you a bit, but at the time I thought it was the end of the world. June Week in the brig! My Annapolis career had begun on the *Reina Mercedes*, and now apparently it was to end there, too. Five of my classmates were also confined—for "Frenching out" or "going over the wall"—but that was small consolation. Only one of them had more demerits than I did, and he was hoping to be tossed out of the Academy anyway and had deliberately *tried* to pile up an impressive total!

For four days, each of them a lifetime, I lived on the ship, forbidden even to look at a newspaper or make a phone call. My parents had arrived and were at a hotel in the town, but I was not allowed to communicate with them. The Academy was in the midst of June Week festivities, but all I saw of the gaiety was a glimpse or two while on the way to make formations—for we had to carry on with the usual routine, of course, even though con-

fined. The difference was that at day's end we were marched
ingloriously back to the brig while the rest of the upper classmen
waltzed merrily off to celebrate their coming graduation.

But the Navy's heart eventually softened. Released from the
ship the night before graduation, I hurried to town, contrite beneath a wise-guy grin, to pick up my folks. Next morning all was
forgiven when I was handed my diploma and commission.

Farewell, Annapolis!

4   *IN QUEST OF WINGS*

I HAD left Annapolis with orders to report aboard the U.S.S.
Nevada, but first there was time for a brief trip home. There I
found an invitation from Emil Shaffner, an old high-school pal, to
attend a Phi Beta Kappa dance at the University of North Carolina. So I really celebrated my "June Week" at Chapel Hill, where
in the white summer uniform of a Navy ensign old Bus Miller
made amends for those days of confinement in the Academy brig.

It was commencement week at the University. Guy Lombardo
was on hand with his orchestra for a series of dances, and the peach
brandy and corn flowed freely in all of the fraternity houses. The
campus football hero, I remember, was prominently on exhibition
at Emil Shaffner's fraternity. A nice boy, too. But the uniform of
the U. S. Navy attracted him greatly, and before long he was
forlornly begging me to swap clothes with him. "Just for tonight,
Bus ol' pal," he teased. "Just for one li'l ol' hour, huh?" It was a
shame, really. But the uniform wouldn't have fit him anyway.

Shaffner decided that night to go to the West Coast with me.
He had a handsome new Buick, a graduation gift from his father.
"Be a chance to break it in," he insisted. "Come on, we'll have a
time."

We did, too. It was the first trip across the country for both of
us and we rubber-necked all the way, stopping at every spot that
could even remotely be called a point of interest. We waved at a

good many girls, too, I remember. At least Emil did. As a graduate of the United States Naval Academy, I was on my dignity. Of course!

When the trip ended I was genuinely sorry and envied Emil his Buick. But the envy soon vanished. At San Pedro lay the battleship *Nevada,* which was to be my home for an indefinite part of the future, and when I first felt her deck under my feet I was a proud young man. There's nothing quite like the feeling of fullness and satisfaction that electrifies a naval officer reporting aboard his first ship. My job as junior engineering officer was not very important (later I became assistant navigator, junior division officer in the antiaircraft battery, and ultimately junior division officer in the Construction and Repair Department), but the *Nevada* was a big, powerful ship, as fine a battlewagon as any in the fleet, and I was part of her. Shaffner could have all the Buicks in Detroit!

I joined the *Nevada* in June, 1931, and promptly made the cross-country mail service work overtime with letters to the girl I'd left behind. Not for long, however. Thelma Davis either tired of reading the letters or took pity on the mail carriers; in August, waving the white flag, she arrived in San Pedro to marry me. Marriage on an ensign's salary, we soon discovered, was not what the lovelorn columns said it was, but at least we saved a lot on postage.

In August there began to be talk among the *Nevada's* junior officers about the elimination flight training program at San Diego. I had no particular desire for a change. Certainly I was not obsessed with any urge to fly. But the talk soon became the central subject of conversation whenever we got together, and someone at last asked, "How about you, Bus?" "Why not?" I said. You lost nothing by trying, and it would be a new and interesting experience. After flunking out, I could return to the ship knowing more about naval aviation and would be a better officer for having made the attempt. By present standards I was pretty old at twenty-three to be starting the pursuit of a pair of wings and thought the rigorous physical examinations would shelve me at the start. Flying is for youngsters.

We endured an endless string of exams. For days, it seemed, we walked about undressed while indifferent medical men poked, pinched and probed us for defects. I passed and was mildly astonished.

The elimination training itself was not too difficult. It was designed simply to weed out those applicants who couldn't fly at all and would obviously not make the grade if sent on to Pensacola. In the course of a month we flew fifteen to twenty hours in training planes with an instructor and then soloed for two hours. (Which brings to mind an item that appeared very much later in the weekly bulletin of Naval Air Station No. 23, just before I went out to the Pacific with Squadron VB-109. "Did you know," the item read, "that our new skipper took elimination flight training at San Diego way back in '31 and *soloed*? Now he flies with ten assistants!")

At any rate I soloed, was given a final physical check-up by Doctor Guy Fish, one of the first of the Navy's flight surgeons, and was suddenly so enthusiastic about the whole thing that I ran practically the whole way home to tell Thelma we were going to Pensacola. (Later I met Doc Fish, then a lieutenant commander, and told him how his stamp of approval had affected me. "Humph," he snorted. "Old men are always tickled to death to be told they're healthy. It's a good thing you didn't run your marathon *before* I looked you over!")

This time I traveled across the country with a wife instead of a high-school pal, and we were in a gay mood. Both of us had been to Florida before and knew we'd like living there. It wasn't until some time later that we acquired the typical Navy family's fondness for the West Coast. At Pensacola we found a pleasant little cottage for rent and I reported to the Naval Air Station for training. It began to look as though Miller was fated to become an aviator in spite of himself.

At Pensacola in those days the training was not highly specialized. Starting with little yellow seaplanes, the student moved upward through five squadrons and after completing the course was considered capable of flying anything on land or sea, including carrier and catapult planes. Squadrons One and Two were the primary course in sea and land planes. If you managed to maneuver through those without too many down-checks, crackups or call-downs, you were reasonably sure of going the rest of the way. After that it became progressively tougher. Squadron Three included scout planes and formation work. Squadron Four introduced the trainee to two-engined seaplanes and single-engined seaplanes on catapults. The last of the squadrons polished him off with some really intense training in fast fighters.

While a student at Pensacola I had one very close brush with eternity. Aloft one day in an F3B, I was making gunnery passes at a towed sleeve (target). Another student, "Swede" Michelet, was up in another Boeing fighter, an F4B, going through the same maneuvers. We were well aware of each other's presence in the sky, of course, but for some reason we suddenly got our signals crossed and sped toward the target at the same time. Michelet made a high side pass at it; I made an overhead pass. With our eyes glued to the gun sights, neither of us was aware of the other's approach, and the two planes converged toward the same point at the same time.

No one could warn us, for we had no radio in any plane. The instructors could only sit there, waiting for us to collide. At that speed, both planes would have come apart like pricked balloons if even the wing tips had touched.

Suddenly in the midst of an almost vertical lunge at the target, I heard a peculiar roar above the thunder of my own engine, glanced to port and saw Michelet's plane almost on top of me. He must have heard or seen me at the same instant, for his eyes were so huge they seemed to fill his goggles. (Mine were undoubtedly as large!) The two planes were hurtling at full throttle, their wing tips only inches apart when we slewed away from each other. I was still in a cold sweat when I landed.

Later I learned another lesson, when the time came for initiation into night flying. For some reason which now doesn't matter, I was asked to go up the road to the Warrington ice-house that evening and bring back a cake of ice. I drove up, stepped out of the car and heard familiar voices, rather loud voices, raised in song. In the dim solitude of the ice-house two of my classmates were holding evening choir practise over a keg of beer which they surreptitiously kept there.

Now there's nothing wrong with a keg of beer in an ice-house, even in Pensacola, nor is there anything sordidly sinful in the sound of two slightly mellow voices raised in song. These boys did not have to fly that night. I did. But I didn't know then—at least, not from personal experience—that flying and alcohol don't blend.

"Have one, Bus!" the boys invited.

Bus had one. One led to another. After several, the remark was made that it was a good thing we didn't have to fly tonight.

"Holy smoke!" I yelled. "I *do* have to fly tonight!"

The boys were sympathetic. They hadn't known. Solemnly they patted me on the back, reassuring me. It was too bad, but perhaps if I were cautious, extremely cautious, nothing awful would happen. A slight crackup, maybe, but nothing more serious than a broken leg or two. With that they returned to their melodies and Miller departed.

"Oh, well," I thought, walking unsteadily in the direction of the Gulf of Mexico a while later, "there'll be an instructor along. He'll do most of the actual flying."

The plane was ready and I climbed into it. The instructor had not arrived. I waited, shaking my head every few minutes to bring the stars into clearer focus. It was a beautiful Florda night and there seemed to be ever so many stars, but I had trouble separating them. Where the devil was the instructor?

He showed up, finally. But before he could reach the plane, one of my classmates hurried over and whispered in my ear. "Bus," he said, "you're in for a tough time of it. Your instructor just drove up with his wife, both of them giggling. He acts sort of woozy, too. Looks as though they've been to a cocktail party, and you're going to have to do the flying tonight."

I didn't have the strength to answer him—just sat there, bug-eyed with alarm, wondering who in heaven's name *would* fly the plane. Then and there I vowed that if I came back alive from that flight I would never again violate the "Bevo List" by taking a drink within twenty-four hours of flying time. As the instructor climbed into the plane, I watched his every movement with a kind of numb fascination. He nodded, and we took off. Not until we had landed, a few hundred years later, did I learn that the whole thing had been a gag and the classmate had been kidding me. The instructor, of course, was not in on it. When I told him the story years later, he was highly amused.

But I was not to be graduated from Pensacola without an accident, and when it happened it was serious. With only a week to go before the end of training—this was in August, 1934—genuine tragedy caught up with me. It cost a man his life.

Seven of us were practising carrier landings one morning at Corry Field, flying F3Bs and F4Bs, the trim little fighters built by Boeing. Carrier pilots of course always begin their training on land, taking off from and coming down on an ordinary airfield marked with chalk-lines or flags to simulate the flight deck of the carrier from which eventually they will operate. The "flight deck" at Corry Field, outlined with flags, was 250 feet long and 70 feet wide. An instructor, representing the carrier's landing signal officer, waved us in one by one as we flew in a landing circle above. Another instructor stood by to observe and correct our mistakes. There was no radio communication in those days between the pilots and field; all signals were visual.

I had made four or five landings O.K. when my turn came up again and the landing officer flagged me in. This approach too was apparently without fault, for he gave me the "cut," which meant that I was O.K. to land the plane and then take off.

The landing signal officer was on my left. I was watching him for instructions. After the "cut," I landed, and, still looking to the left, gunned the engine again. For some unexplainable reason I made a low, fast take-off, holding the plane just off the ground until I had about eighty knots.

Suddenly, with the wheels about three feet off the ground, the right side of my plane seemed to hit a brick wall. It shuddered as though coming apart, and I heard a ragged, ripping sound above the noise of the engine. Instinctively I closed the throttle, squeezed down in the cockpit, and hung on.

The plane wobbled crazily, swerved to the right and struck the field, skidding on its belly and plowing up the turf in a deep, crooked furrow. When it stopped I climbed out, physically unhurt except for a shaking up; but mentally muddled and uncertain which way to turn. The plane had been swung around almost 180 degrees. Back up the field, where the plane had hit something, I saw a mowing-machine—one of the tractor-drawn mowers used by the maintenance crew to keep the grass trimmed on the airfield. Its canopy was gone. The driver had slumped backward on the seat, and I knew from the limpness of his body that he was hurt. At the moment, that was all I knew.

I ran to him—ran all the way back, with the parachute thumping my legs until I unhooked and dropped it. The field seemed miles long; I thought I would never reach him to lift him into a normal position. He had fallen from the iron seat of the tractor, I saw then, and caught his foot in the pedals. His head was bleeding.

Others reached him just after I did, and a lieutenant helped me to extricate him and lay him on the grass. He was badly hurt— very badly hurt. The right wing and landing gear of my plane had struck the tractor at better than eighty miles an hour, shearing the canopy off. The plane's propeller had struck the right side of the man's head. The landing gear, torn loose by the impact, had been flung the width of the field.

I was white and shaking, and desperately I wondered why the signal officer had given me the "cut." For with my attention fixed on him, I had not even seen the tractor before I hit it. (Later we learned that, due to the rise in the field necessary for drainage, the landing signal officer himself could not have seen the tractor.)

Methany—that was the man's name—was rushed to the hospital in an ambulance, and after I had been questioned and had written a statement of my version of the accident, the hardest thing I've ever had to write, I went to see him. There was not much hope, the doctors said. On the way home I kept repeating, "Dear God, don't let him die. Dear God, please don't let him die."

After talking to Thelma, telling her what had happened, I drove out to the man's home, following directions given me at the field. It was a hard place to find, located some fifteen miles out of Pensacola on a dirt road that seemed to twist and crawl forever through scrub palmetto, "black jacks," and Florida desolation. I had been told to look for an old frame house set back from the road a bit, a poor place with no electric lights, no conveniences. When I found it there were children playing in the dirt yard—four of them, and all of them small. Mrs. Methany was expecting another, soon. I thought I would never again have the courage to step into an airplane. I thought I would never want to.

For two weeks he clung to life. Every day I drove out to the old frame house on the dirt road and took his wife to the hospital to visit him, and she sat by his bed. The doctors told me there was no hope, no hope at all. But they did not tell Mrs. Methany that, and day after day she sat there waiting for him to regain consciousness. He never did. He just lay there.

I took her to a doctor, too, for prenatal care—the first she had ever had—and paid in advance for delivery of her baby. And one day while returning from such a trip she mentioned her mother in Alabama and expressed the wish to go there. So with the four youngsters in the back seat, we went. It was a very long way. The kids tired of the ride and began whimpering, then fought among themselves. I wondered desperately how she would ever bring them up, and what she would do with them when the new one came—for we had two children of our own now and I was acutely aware of the responsibilities that went with them. On that long ride to Alabama I knew that I would never forget what I had done to Methany. I never have.

Her mother's house was as humble as her own, situated in Alabama back country some distance from the main road. We were there for hours, and the older woman kept staring at me with her small, intense eyes, until I felt like a man who had been branded for some evil deed. I knew she hated me. Whether she blamed me or not for what had happened, I had been the instrument of Methany's misfortune and for that she despised me. When I got home that night I hardly had the strength to walk into the house and throw myself down.

Meanwhile, orders had been handed me to report to the Asiatic Station, but of course I could not obey them. A court of inquiry had been formed to investigate the accident; as defendant I could not leave the station. Meanwhile, too, the course at Pensacola came to an end and with twenty-two others of the original class of forty-five I was graduated. The day I received my wings I went with Mrs. Methany and the children to the hospital, as I had been doing right along. Methany had never regained consciousness. That day he died.

Captain Halsey—the same Captain Halsey who had commanded the *Reina Mercedes* at the Naval Academy—was head of the court of inquiry, and at first I was the sole defendant. For days and days the hearing dragged on. I had a beaten-down, hopeless feeling that some sort of bad luck was hounding me. My mother had died soon after my enlistment in the Navy. My Annapolis graduation had nearly been canceled by demerits. Now this.

Eventually, however, the landing signal officer, who had been on the field that day, and a chief petty officer were named defendants with me, and the investigation brought to light a curious fact. It went something like this, when one of the maintenance men was called before the court and answered questions:

"Did you tell anyone you intended to run the mowers that morning?"

"Yes, sir, we told the chief here and he said he would inform the squadron duty officer."

"Chief ——, is that correct?"

"Yes, sir. One of the men did tell me."

"At what time?"

"About eight o'clock."

"Did you inform the duty officer?"

"No, sir. I meant to, but someone said I was scheduled to put in some flight time. I didn't contact the duty officer."

"Then the students and instructors on the field were not informed that the mowers would be in operation?"

"No, sir, not by me."

The investigation took a new turn after this, and I was absolved of blame. Naturally I felt very much better about the whole affair. But I was still responsible—a pilot is always solely responsible for making sure that his flight path is clear. Methany was dead, his kids were fatherless, and the wings I wore seemed not very bright.

A few days later I was ordered to report aboard the U.S.S. *Ranger*, at Norfolk, for duty involving flying in Bombing Squadron Three.

5   *CARRIERS AND CRUISERS*

THE *Ranger* was a new ship when I joined her. Commissioned in June, 1934, she had just returned from a long shakedown cruise to South America. As carriers go—even as carriers went then—she was not a big ship; the *Saratoga* and *Lexington*, commissioned seven years before, were both larger. The *Ranger*, however, was the first American ship designed from the beginning as an aircraft carrier. At Norfolk she was awaiting the arrival of a newly formed dive-bomber squadron, VB-3, and my first job was to go to Cleveland with several other pilots, pick up our BG (Great Lakes) planes and deliver them to the ship. On the way north from Florida I dropped Thelma and the youngsters in Winston-Salem where they rented the very same apartment in which Thelma had been living when I had phoned her that evening, long ago, to patch up the quarrel over my enlistment.

After flying the planes of VB-3 safely from Cleveland to Norfolk, we had trouble with them at our destination. Circling above the ship, not at all sure of our ability to land on her (this was the first actual carrier landing for most of us), we anxiously awaited instructions. At the come-in signal the first plane hit the deck, slewed along it and crashed into the barrier.

I steadied my nerves with a deep breath and tried to review mentally, in the space of a few minutes, all that I'd been taught. This was the first ship landing I'd ever even seen, and the initiation was hardly reassuring!

We continued circling while the wrecked plane was pulled aside and the barrier repaired, and then our number two man went in. His approach was faultless, he seemed to cut his throttle at exactly the right instant, but he, too, plunged into the barrier. This time, however, we saw what the trouble was. The planes' hooks, supposed to drop down and catch the arresting wires stretched across the flight-deck, had bounced up again—and stayed up—on contact with the deck. Both planes had missed all the wires and continued unchecked until the barrier halted them. Obviously something was wrong with the hooks on the planes we were flying.

Rather than risk further accidents we turned about and flew back to Norfolk, landing on a field there. Later we taxied the planes through the streets of the Air and Training Stations, past hundreds of gawking spectators, and were hoisted aboard the ship by crane. It was hardly the way a flock of bright new Navy pigeons should have been delivered to their cote, and for the chagrined pilots the trip through the station streets was deflating, to say the least, but the planes did eventually reach their destination. A few days later, with the hooks repaired, we began to fly in earnest.

Those were hectic days. Carriers, remember, were not then the proud first-line ships they are today, and naval aviation was still looked upon in some quarters as an awkward step-child. Ships, planes and techniques were still in the experimental stage, subject to everlasting revision. In the entire fleet we had but four carriers: *Langley*, 1922, (converted from the collier *Jupiter*), *Saratoga* and *Lexington*, 1927, (sister ships originally laid down as battle cruisers), and *Ranger*. The number of pilots qualified to serve these ships was so small that today it would seem ridiculous. Crackups were common. As Commander Bob Winston says, you could usually recognize a carrier pilot by the marks of the instrument panel impressed upon his face by too many sudden decelerations.

We had some close calls on those early practise flights—so many, in fact, that aboard the *Ranger* we were ordered to leave our mechanics on the ship and ride with sandbags in the rear seat instead, because good mechanics were hard to replace. Sandbags were cheap.

It was a sandbag, however, that nearly cost "Dub" Johnson, one of VB-3's fliers, his life. Dub was practising take-offs and landings one day with a bag in the rear cockpit when suddenly, while he was aloft, the bag shifted and jammed his rear controls. He couldn't reach it, of course, to free the controls, nor could he attempt a landing on anything as short and narrow as the *Ranger's* flight deck. Luckily we were not far from shore, and he was able to find a field, fairly level, a short distance from the beach. But he barely made it. His wheels stripped leaves from the trees as he sweated the plane in for a landing. (Dub was shot down later in one of the first raids on the Marshalls and when last heard from on the radio was a prisoner of the Japs.)

Another time on the *Ranger* a classmate of mine from Pensacola, "Mandy" Ryon, coaxed a plane back to the ship on what he thought was his last drop of gas, only to find when he landed that his right tank was actually full. A plane of this type is equipped with three tanks, right, left and main, plus a small sump tank which holds five gallons. The fuel is pumped to the engine from whichever tank is indicated by the fuel selector. Investigation disclosed that Ryon's fuel selector was 180 degrees off—in other words, while the indicator pointed to the right tank, which was full, he had really been trying to draw fuel from the left tank, which was empty. But for the cupful of gas in the sump tank that finally brought him home, he might have crashed into the sea without ever knowing what was wrong.

We talked over these misses and near misses in the ready room, not through any liking for such discussions but because it paid off. No pilot enjoys a recital of the accidents that befall his mates—especially the ones with unhappy endings—but every scrap of information is likely someday to be of value. Had we not listened to Mandy Ryon's story of the fuel selector, I myself might have failed to come back a few days later, for the same thing happened to me.

One day we took part with the *Langley*, *Lexington* and *Saratoga* in a fleet exercise off the Panama Canal. Two of the carriers were to launch an attack against the Canal Zone; the other two were to fight the assault off. To equalize the opposing forces, aviators from the ships were shuffled about somewhat, and I was assigned to the *Lexington*, to "defend."

The *Lex* was blessed with some air officers then whose ideas on naval aviation were well ahead of the times. They went into a huddle. The way to knock out an enemy carrier, they decided, was to catch it with its pants down—that is, by surprise, before its planes could take to the air. In the ready room we were briefed for an early morning surprise attack, and before dawn we were off on the mission. Scouts departed first, to hunt out the *Saratoga* and radio her position. The fighters and dive-bombers followed.

The attack was completely successful. We found the *Saratoga* and theoretically dive-bombed the daylights out of her before she knew what day it was. Hardly a plane left her deck. Then, smug and self-satisfied, we turned for home. The *Lex* of course was maintaining radio silence lest she too be caught unprotected and "sunk," but we knew where she was. Hadn't we flown from her deck just a short while ago?

Only she wasn't there—carriers are almost never where you expect them to be!—and the area in which she should have been cruising while awaiting our return was heavily overcast. We flew back and forth through the cloud layers, anxiously searching. Our ship had vanished. The soup became thicker and thicker until the sea was blotted out for miles.

I could see the other two planes of my section; in fact, with an unholy fear of getting lost I stuck so close to Bob Brixner, our section leader, that he must have thought I was trying to hitch a ride home on his tail. The number three man, Lorry Forbes, was taking no chances either and flew so close that I could see him shaking his head as he peered down through the weather in search of the ocean that wasn't there any more.

As the soup thickened and our gas supply diminished, I worked on the mixture control to save what little gas was left and saw Brixner, with his headphones on, angrily cussing because the ship would not heed the captain's call for radio aid. Finally we knew we were lost—we three and all the rest of the planes of our three squadrons. Scouts, fighters and dive-bombers were all milling about above the overcast, frantically seeking a place to land. Brixner and his mech began peeling off their clothes to ready themselves for a struggle in the water. Forbes took the cue and stripped off his helmet. So did I—and I began thinking of some of the stories I'd heard about sail-fishing in southern waters, how a hooked fish would sometimes be ripped to pieces by sharks before it could be brought to gaff. Our "victory" over the *Saratoga* was going to have a hollow ending.

Then I saw smoke—a lovely gray plume of it poking through the overcast to beckon us. The admiral, realizing our predicament, had called off the exercises and ordered the ships to make smoke to guide us in.

The decision came not a moment too soon for some of us. Following the smoke, we let down through the overcast and at last found the ship, a ghostly shape on a gray ocean, awaiting our return. As we circled for instructions she radioed an anxious query: "Who is lowest on gas?" She got a prompt and frantic reply from one pilot who made a bee-line for the deck without any further hesitation. He didn't circle for position; he just went in, bounced over the stern ramp, hooked the wire and stopped. There was not enough fuel left in his tanks to taxi him off the wire!

One by one we circled and landed until at last all were aboard. No one in our squadron had failed to come back; there were no accidents. Three planes of our fighter squadron, however, had gone down in the drink—the pilots were rescued—and in the ready room, afterward, some of the boys solemnly plucked new gray hairs from their heads while awaiting interrogation on the results of our mission. We had knocked out the *Saratoga*, to be sure, but that was make-believe stuff. There'd been no masquerade about our own near catastrophe!

As a carrier pilot I served aboard the *Ranger* and *Lexington* in VB-3 until June, 1936, and then was transferred to the heavy cruiser *Louisville* as gunnery, navigation and material officer. The

Louisville was commanded then by Captain (now Admiral) W. S. Farber, who had been prominent in the development of the catapult. (With the *Enterprise, Yorktown, Chester, Northampton, Salt Lake City, St. Louis* and escorting destroyers, the *Louisville* took part in the first naval assault on the Marshall and Gilbert Islands, in February, 1942—a region I was destined to know rather well, later, as commanding officer of a Liberator squadron.)

Nothing much happened to me aboard the *Louisville*, but one day I discovered a new wrinkle in naval aviation by landing an SOC —a single-engined float plane built by Curtiss—at full throttle. That little stunt hadn't been done before, I'm sure. Probably it hasn't been done too many times since. Despite my demonstration, the manual was not rewritten. This is what happened:

We were having routine anti-aircraft practise off Lahina Roads, in Hawaii, one rather gusty day with the sea running rough. The plane on the port catapult was shot off to tow a target, but the target carried away when it was streamed, and the pilot returned to the ship. With a lad named Foster, my radioman, I climbed into the SOC on the other catapult. A new target with fifteen hundred feet of manila line was fastened in place on the starboard wing, and away we went.

The thrust of the catapult, however, jarred the target loose and it streamed out behind us as we cleared the ship. We hadn't sufficient altitude to keep it off the water. Fluttering on about a thousand feet of limp line, it splashed into the sea, setting up such a drag that the little plane nearly came apart.

With full gun, squeezing every bit of speed out of the engine, I fought for altitude and yelled at Foster to "Release it! Release it!" But he couldn't. The line had fouled on the wing and was caught fast, tending to slew me and spin the plane to the right.

We never did gain more than two hundred feet of altitude, even with the SOC shivering her heart out in the attempt. The pull of all that heavy manila line was too great. About half a mile from the ship the plane gave up the battle and dropped like a stone, striking the sea with full throttle and barely above stalling speed. The tow-line broke, I cut the throttle, and we bounced. It was like skipping a stone. But an SOC is a lot more fragile than a stone, and why she didn't collapse like an accordion I don't know.

When we stopped at last, Foster crawled out on the wing to assess the damage we had sustained. The float of course had absorbed most of the shock. It looked like a wash-board. The smaller wing-tip floats were wash-boarded, too, but less seriously. Apparently nothing much else had suffered. With a look of incredulity on his face Foster said, "Darned if she isn't all here!" So I taxied gingerly back to the ship, still expecting us to come apart any minute, and the plane was lifted aboard.

An incident occurred then that will illustrate how very far naval aviation has advanced since the days of the first carriers and catapults. On the carriers the airmen were in their own element, of course, but in those days the men who flew from catapults on battleships and cruisers sometimes had their troubles. It's no secret now that the fliers assigned to these ships were not always welcomed with brotherly love by the regular ships' companies. The extra flight-pay granted to aviators had something to do with this cool reception, I suspect, and so, undoubtedly, did the cockiness of some of the aviators themselves. At any rate, friction sometimes existed.

When the damaged SOC was hoisted aboard the *Louisville* and placed on the catapult, both Foster and I were understandably a bit woozy. We'd had a severe shaking up and a thundering good scare. Not too steady on our pins, we clambered out of the plane into a circle of interested but silent spectators. I glanced at Foster and he shrugged. Ignoring them, we turned to inspect the plane. Apparently no one was even glad to see us safely aboard.

Just then a messenger arrived from the bridge, pushed through the crowd and smartly stepped up to me. "Sir," he said, "the Officer of the Deck wishes to remind Mr. Miller that he is ten minutes late relieving him on watch!"

Too stunned to answer, I stood there with my mouth open. The Officer of the Deck had certainly seen the accident. He knew why I was late relieving him. He knew that only by the grace of God and a few good gremlins had I escaped reporting to St. Peter. And now Mr. Miller was ten minutes late!

I went to my room, changed into whites and climbed to the bridge where the OOD was waiting. The smug smile on his face was a give-away. Here at last was a chance to call down a despised aviator, and he meant to use it to the utmost. I think I might have done something about the smugness of his smile if Captain Farber had not stepped from the chart house at that moment.

The captain stopped short in surprise when he saw me there. "What the devil are you doing here, Miller?" he demanded.

"I have the deck, Captain."

He swung abruptly to the OOD, gave him a look that removed the smug smile completely, and angrily ordered him to call the senior watch officer. When the latter arrived, the skipper really exploded. "If we have any more of this sort of thing," he snapped, "we'll take the aviators off the watch list altogether! Now, dammit, find a relief for Miller!" Then, his anger fading, he turned again to me. "Son, I'm glad you weren't hurt," he said. "Better go to your room and take things easy for a while."

Those were the days! But I didn't go to my room. Foster and I returned to the plane and went to work. With the rest of our aviation gang, we removed the damaged float and installed a new one— a mean job with the ship rolling in a rough sea—and finished, dead tired, at eleven o'clock that night. And despite the clash with the OOD I was in good spirits, for we were homeward bound with every prospect of a long leave, and I had not seen Thelma and the youngsters in a long time.

The hoped-for leave did not come off. En route to the West Coast from Hawaii the *Louisville* picked up an SOS from the British motor ship *Silver Larch*, which was on fire with a cargo of naphtha about eight hundred miles from our position. Speeding to help her, we steamed under forced draft through dirty weather all night and reached her at ten the next morning. Her crew of thirty were struggling to control a blaze which flared up anew each time the hatches were opened. Her skipper wondered if, despite the high

wind and extremely rough sea, we could take off her eight passengers.

Captain Farber decided to try, and asked for a volunteer boat
crew. They were brave boys, especially the boatswain in charge.
We watched the little boat fumbling through the storm on the way
back with its load of passengers, saw it tossed and pommeled by
the waves until it seemed certain to founder at any minute. The
Louisville herself was taking a pounding. But the crowded little
boat made it, and when she came alongside we plucked her from
the sea, occupants and all, and hauled her up with a last-ditch heave.

To the half-drowned people from the *Silver Larch* we were
heroes then. But, dammit, we had to take them to Hawaii—and
every man aboard the cruiser had been eagerly counting off the
miles to San Pedro! To make matters worse we reached Pearl
Harbor on a Sunday, when all facilities for refueling the ship were
shut down. On Monday, at last, we started home. The *Larch*, her
fires in check, was towed to port by a Coast Guard cutter.

The prospect of getting home was especially important to me,
for I had been expecting news of an arrival in the Miller household
at Long Beach. As we made for San Pedro I haunted the radio
room, hoping for a message. But rumors suddenly reared their nasty
little heads. Perhaps we weren't going to San Pedro after all. The
radio room had picked up reports that a number of naval vessels
were being sent to the South Pacific to search for a plane which
had disappeared there on a flight from New Guinea to Howland
Island. Amelia Earhart and Captain Fred Noonan were missing.

We might have been assigned to the search, but the old *Louisville*, just to assert her independence, burned out tubes in two of
her boilers instead. With the engineers grumbling below decks to
keep us crawling, we finally limped into San Pedro many days behind schedule. Leaves were canceled. Babies and other unimportant
matters would have to wait. We steamed sullenly up the coast to
Bremerton for repairs and then turned into the Pacific again for
fleet maneuvers.

Soon after that the expected addition to the Miller household
arrived and I received word in the Pacific that mother and Sally

were doing nicely. But I still had little love for the *Louisville* at that
time!

Eventually, however, my assignment at sea ended with orders
to report to Pensacola again as an instructor. The change was an
important one. It marked the termination of three years of service
aboard the *Ranger*, *Lexington* and *Louisville* and the beginning of
a life ashore.

The three years at sea had been exciting, for even in those peace-
ful days the Navy was a most active organization, constantly experi-
menting with new ideas and equipment. By now I had prowled a
good part of the fleet's stamping-ground, had flown dive-bombers
and scout planes over most of the Atlantic and a large part of the
Pacific. Maturing, I had begun to pick up some of the sound Navy
knowledge that was to come in so handy later on.

6   *PENSACOLA PARAGRAPHS*

FLIGHT instructing is a peculiar and sometimes frightening business.
You are dealing, first of all, not with a youngster of average intel-
lect and initiative but with a student who by virtue of having passed
certain stiff tests and examinations has proved himself a little better
than run-of-the-mill. Frequently you learn as much from him as he
does from you.

I had forty-one students at Pensacola and soon learned not to
form snap judgments of their ability. Some, of course, were "hot
shots." In any group of that size you are likely to encounter two
or three who think they acquired, at birth, all the little tricks and
techniques which other pilots must learn through hours of patient
practise. At the other end of the group will be a few whose sluggish
reactions and apparent inability to think through even the simplest
problems will cause you to wonder how they ever passed the elimination flight tests. In the beginning, therefore, it was dangerously
easy to form opinions. Then the hot shots pooped out; the slow,
hard-working boys retained what they painstakingly learned and

became good, steady pilots, and your opinions had to be revised. Not always, of course, but often.

The students had been subjected to some primary seaplane training and had soloed once before I took them over. My job was to teach them takeoffs, landings, and straight flying at first. I had been through the same kind of training when a student myself and consequently knew, or thought I knew, pretty much how they would behave.

I soon learned that all men do not react alike, and not even a very reliable student will respond the same way day after day. Little things make a difference. If a boy happens to be nursing a cold or a headache, if he has been out late the night before or has just received a disturbing letter from home, he may do something entirely unpredictable. In short, he is a human being. So he must be made to perform the same tasks over and over again, despite the monotony of such training, until no matter what his personal approach to the problem, he will automatically go through the proper motions. Needless to say, that sort of thing becomes as tiresome to the instructor as to the student. We welcomed any diversion.

Doc Defoney provided a suitable diversion for a time. Doc was one of the first of the Navy's flight surgeons, and for a long while the instructors at Pensacola had been patiently praying for a chance to even the score with him. Not that he was a tyrant. Not at all. He was a grand guy and a fine surgeon. But at forty he had seen a lot of pilots come and go, and most of us had squirmed under his good-natured probing.

Doc's sense of humor was at times devastating. "Headache?" he'd say. "Huh. Nothing much there to ache, I'd say. You'll live, I expect. Hope not, sake of the Navy, but guess you will. Better keep out of those flying machines for a while, though. Kill yourself, otherwise." His "psychos" or psychological exams were famous among naval aviators and usually included such ribbing questions as "When did you quit sucking your thumb?"

Doc kept a lot of us alive, I suppose, but he kept alive, too, his own reputation for being a very funny fellow, always kidding, ha ha. The boys longed for an opportunity to do a little counter-kidding. Trouble was, he outranked us.

Then one day he decided he ought to take flight training. "Can't be a real good flight surgeon without knowing how to pilot one of those contraptions," he said. "Got to know what goes on inside the minds of you fellows—if anything." Despite his age, he wangled permission from the front office, and then—ha ha—he was at our mercy. Because we, whom he had poked and probed and kidded for so long, were now his instructors!

Doc was assigned to me as a student, and when it came time for stunting he was jittery. At his age he had a right to be. Immelmans, loops, falling leafs, snap rolls, split S's, cartwheels, spins and such maneuvers are not for forty-year-old stomachs. Besides, an older man is likely to be somewhat slow in his reactions and, sensing this, may be dangerously cautious. An over-cautious man at the controls of an airplane can often get into as much trouble as a hot-shot pilot who thinks he is afraid of nothing at all—just as the man who never drives his car faster than twenty miles an hour sometimes invites a smash-up on the highway through sheer timidity. Doc knew this of course, and was worried.

As a matter of fact, almost any pilot dreads his first stunt check, anticipating the worst and not knowing how under the sun he may react. You simply don't know what will happen, and the fact that you loved roller-coasters as a kid is no indication that a spin won't separate you from your breakfast. (As a kid I never could stand roller-coasters; but stunting in an airplane doesn't bother me at all.)

Doc's particular bogey, it developed, was the split S, which for that matter is no lark for anyone. Aloft one day, I put him through his paces and then matter-of-factly suggested a split S or two before heading homeward. Doc nodded rather grimly and executed the maneuver nicely. But when I looked up into the mirror to see how he was doing, his face was the color of old milk and he was gasping. He told me later that he suffered from sinus trouble and the split S caused the fluid to drain into his mouth where he had to swallow it or choke. But even that didn't stop him. He was a very stubborn guy and a very grand one. "Come on, Commander," I used to say when the stunting began to wobble him. "No use punishing yourself. Let's go home." Later, when we had landed, he would snort at me and say, "Lieutenant, I came here to fly, and I'm flying!" Then he would have to run the gauntlet and take his ribbing from former victims. "Doc," they'd say, "you don't look just right. Better report for a check-up right away." Or, "Doc, you'll be grounded for good this time, sure. Sake of the Navy, Doc." But Doc had the final laugh, after all. He passed his flight training and moved right back into the driver's seat, and then— ha ha—his sense of humor knew no bounds.

Speaking of watching a student in the mirror, I learned rather early in the game not to place too much faith in what the mirror revealed, and toward the end I stopped using it entirely. A man's face is often misleading, and a confident, smiling student can crack up a plane quite as easily as one who appears to be emotionally in a dither. For example, the boys had to be taught how to land on small emergency fields, with or without use of throttle. As I flew along with a student, I would search the ground below for a likely spot—usually a cow-pasture—and then cut the throttle. It was up to the student then to slip the plane into the pasture without killing us.

One day, having spotted a fair-sized field in the midst of an area sprinkled with blackjack oaks, I went through this procedure while watching the student in the mirror to observe his reaction. He grinned and I relaxed. He was still grinning when we sliced the tops off a pair of oaks and hit the ground with a wallop that nearly bounced me out of the cockpit! Never again did I place any faith in a facial expression.

The planes used by the students took quite a beating, of course, and had to be checked often. For this we had five test pilots, of whom I was one. It kept us busy in our spare time, and I personally made over twelve hundred test hops, just to be sure that when some young fellow went aloft to learn to fly, his plane would at least cooperate by doing what he asked it to do. To guide us in checking the planes, the student pilots had what we called "flight inspection forms" or, more familiarly, "yellow sheets," on which they marked down any deficiencies they discovered. Some ailments, of course, were imaginary, and some of the complaints were hilarious. One earnest young man, reporting trouble with the oleo fluid in his landing gear, wrote: "Oleo margarine is leaking out of the right strut." Another complained that his plane flew "with its right wing too low." But we checked them all. Better a dozen false alarms than one fire.

One night—and you may remember this, for it was headlined across the country—some advanced students were returning to Pensacola from a night cross-country hop when the weather turned bad and a heavy fog rolled in to blanket the air station. I was duty officer at Corry Field that night, and as the fog thickened I began to be anxious. Those kids were fairly competent—they had made night landings before—but I could scarcely see the field myself. Long before we heard the returning planes, the fog had become a solid gray shroud enveloping the whole countryside.

They circled, low on gas after their long flight, while we frantically trucked in a huge Army searchlight and set it up at one end of the field to guide them. But they couldn't see the searchlight. Around and around they flew, helplessly seeking a hole in the fog-layer. Even the sound of their engines was muffled.

"We can't see a thing, not a thing," the instructors told us over the radio. The students had no radio in their single-seat fighters. We on the ground could do nothing.

In the end all but one of them flew away to attempt landings elsewhere—if they could reach an "elsewhere" on their dwindling gas supply. The one who stayed was a student from South America who had done a good deal of flying before coming to Pensacola and was sure he could safely bring his plane in on instruments.

He could faintly see the glare of the searchlight, and started a perfect instrument approach on it. We heard him coming in. The several thousand people on the field were very still, listening, almost not breathing.

He almost made it. Ground officers stationed at that end of the field said later that he emerged from the fog in a good approach, about fifty feet off the ground, apparently in full control of his plane. As they sweated him in, it seemed certain he would land without mishap. But suddenly the plane wobbled.

Discussing it later, we came to the conclusion that the searchlight must have bewildered him. It had been set up with the beam pointing skyward, to provide a link between the lost pilots and the field—just as the carriers had made smoke to guide their fliers down through the overcast during fleet maneuvers off the Canal. But evidently—so we reasoned—this man had anticipated a beam of light reaching horizontally along the field, and when he flew over it he made the fatal mistake of thinking his instruments were wrong. He apparently thought his wings were vertical when, of course, they should have been horizontal. The plane suddenly turned sideways, raked the ground with a wing tip, and crashed. The pilot was killed.

One other pilot was killed that night when he failed to bail out as his plane went into a dive. Four men successfully landed their planes on a field north of us. Six others took to their parachutes when their fuel ran out, and after some hair-raising experiences managed to notify Pensacola what had happened to them. One landed in neck-deep water on a shoal in Escambia Bay where he stayed until rescued hours later. As for me, I spent the rest of the night and most of the following day rounding the fliers up and trying to locate the wreckage of their planes.

One day while instructing at Pensacola I took up an N3N, a little seaplane made by the Naval Aircraft Factory and equipped with a Wright Whirlwind engine. The N3N was not exactly a popular airplane then; there had been a number of stubborn "bugs" in it that defied correction, and it had bad spin characteristics. My assignment on this occasion, with an instructor of Squadron One as passenger, was to investigate and report on the spin characteristics of this particular plane in a test over the Gulf. In other words, I was to put the ship into a spin, bring her out of it, and then compose a detailed report on both her behavior and mine.

At 5,000 feet, which should have been altitude enough, I put the plane into a spin as instructed—though with more than the usual apprehension, for I'd flown this same plane before as a land plane and found it hard to handle. We went into the spin with remarkable ease; if I do say so, it was an elegant performance. With a touch of pride I applied stick and rudder to pull us out.

We didn't pull out. The N3N hiked her nose up, her engine went into deep silence, her prop kicked backward before I could catch it with the throttle, and we were suddenly in a frightening *flat* spin without power, fluttering through space like a leaf on a windless day. A normal spin is one thing; the flat variety is something else again and infinitely less pleasing to the insurance people.

In the N3N land plane I had uncovered some peculiarities in the sequence of applying rudder and down elevator. In the seaplane these were accentuated, and I didn't know what to do to straighten us out. It was a problem that called for a prompt decision, but I couldn't make up my mind whether to apply the rudder first or the stick—in fact, was afraid to touch either. "Maybe you ought to do both at once," I thought. "Or will that just wind us up tighter?"

It wasn't funny. We were spiraling toward the Gulf at fearful speed and I was shaken by the panicky feeling that comes to any pilot who has lost faith in his plane. I didn't think we were coming out of that spin. And because I was scared, I did the only thing that could have saved us—hung onto the controls as I had done in a land plane.

Six or seven times, violently, the N3N spun. Then suddenly her nose went down, her prop began turning and her engine caught. With power I straightened her out and woozily headed for home.

It had been a close call, no foolin'. Not for some time did I stop shaking and recover composure enough to think of my passenger. Then I looked to see how he was getting along.

He wasn't interested. For the first time in his life he was deathly air-sick!

Later—partially, I believe, on the basis of the report turned in on this flight—the Bureau of Aeronautics rewrote its technical order on recovery from spins. I remember reading the order and finding it faintly amusing. The pilot was instructed to "stick by the controls despite the turns" if his plane went into a spin without sufficient altitude. They probably don't know to this day that Miller only stuck by those controls because he didn't know what the hell else to do!

Yes, instructing had its moments, though most of it was understandably a grind. Compensation lay in the fact that from nearly every student I attempted to teach, I learned something in return. In combat with the enemy, some of my former pupils have chalked up excellent records. Of that I'm proud.

7   *NEUTRALITY PATROL*

WHEN I left Pensacola in June, 1939, to join the PBY squadron known as VP-11, I reverted abruptly from teacher to pupil again. I had never flown one of these big twin-engined boats and had much to learn about them.

My instructor was a likeable young cadet named Ted Royer, who at first felt excessively rank-conscious (I was a senior grade lieutenant) and refused to accept my professions of ignorance. He soon discovered that I was not merely being modest, however, and after a few days and nights of practise in the essentials, especially in beach, buoy and ramp approaches, I qualified and was made patrol plane commander. Then began the usual peacetime training grind—night flying, formation flying, bombing, gunnery and navigation practise, classroom work in tactics, communications, the use of radio, and so forth. Our operational base was Norfolk.

One morning—it was September 1, 1939, and VP-11 had by then become VP-54—six of our planes were turning up on the ramp in preparation for a formation hop when the leading chief came running from the flight office to give us the cut. "Hold it," he said excitedly. "We've just had word that Germany invaded Poland this morning! All hands are to stand by for further orders!"

We climbed out and stood there in a kind of vacuum for a moment, wondering what to do. Then, realizing the significance of the chief's statement, we rushed to the flight office to get the dope.

Nothing much was known then, of course. The information on hand was barely enough to provide fuel for speculation. Yet some of us, at least, sensed that we were standing by for news of an event which was to turn the world inside out, and we were impatient. "Why in hell," we griped, "doesn't someone tell us something?" No one did because no one could. And, of course, no one did anything.

The following day, however, someone in Washington quite decisively did something, and four of our twelve search planes took off for Newport. Leader of the group was our skipper, Buster Isbell, the same Captain A. J. Isbell who later distinguished himself as skipper of the escort aircraft carrier *Card*, which won the Presidential Unit Citation for outstanding action against German submarines in the Atlantic. The other pilots were Johnny Gannon, Justin Miller and I. Overnight the United States had entered into a "state of preparation," and we were part of the first Neutrality Patrol, set up to watch the Atlantic for signs of hostile activity on the part of America's potential enemies. (Considering the volume of work involved in organizing such a patrol, the speed with which we became operational would seem to indicate that the Navy was neither as sleepy nor as impotent as some of its more voluble critics persist in exclaiming.)

Johnny Gannon and I flew due east from the Virginia Capes for one hundred miles and then straight on up to Narragansett Bay that day, and anchored in Potter's Cove. By nightfall we had found living quarters for ourselves and our crews in a house on Gould Island. Captain Isbell and Justin Miller traveled to the same destination by way of Newfoundland, sizing up the shipping situation to the north of us. Next day two more PBYs of the squadron arrived and the six of us began a routine search schedule. The remaining six planes of the squadron stayed behind to work out of Norfolk. That way we were able to patrol a good part of the Atlantic every day.

On October 2, the Congress of American Republics, meeting at Panama, reached an agreement concerning the establishment of a neutral zone surrounding the Americas (excluding Canada) and raiders and submarines of belligerent nations were forbidden to operate within a belt extending roughly three hundred miles out from the coast. The Navy hurriedly withdrew a number of decommissioned destroyers from the "inactive" pool and recommissioned them for patrol duty, and our task was to cooperate with them in patrolling the neutral zone. Specifically our task was to report the name, size, speed and course of every ship we encountered, and especially to report any vessels whose behavior aroused our suspicion.

It was a tall job. The United States, remember, was still doing business on a cash-and-carry basis with both Britain and Germany—though the British Navy saw to it that most of this commerce was done with Britain. Before long the ocean was literally covered with ships, and one day I reported fifty-four in my search area alone. As for peculiar behavior, we met with a lot of it, but since most of the ships were British the peculiarities were not usually deliberate.

Once, for instance, the squadron received reports of furtive activity at night off George's Bank, some 200 miles east of Cape Cod. Signals and flares had been seen out there, indicating possible collusion between submarines and sympathetic surface ships. One suspicious tanker had been seen regularly in the vicinity. "Check up on this," we were directed, "and keep an eye out for that tanker."

Extending my patrol to include the area one afternoon, I came upon a tanker, flew low over her, and saw that her stack insignia had been painted out and she bore no name. I signaled her, demanding identification. She refused to answer. Certain then that she was up to no good, I zoomed her and saw uniformed men scurrying along her deck in an effort to escape detection. I zoomed her again and saw behind the pilot house a sign bearing the ship's name—obviously the sign had been removed and hidden there. When I reported her, the Navy dispatched a destroyer to investigate. She turned out to be British. I never did find the submarine that was supposed to be lurking out there, and no one ever bothered to tell me why the tanker behaved so strangely. Possibly she'd had a brush with the enemy and was scared.

That was how we worked, patrolling the Atlantic day after day, out of Newport or Norfolk, reporting everything we saw. In the beginning we carried water-filled bombs, four under each wing. They resembled the real thing, and some of the ships over which we flew probably thought we were lethal enough to make things hot for them. Actually they were good for nothing but practise bombing, and when the supply was not too low we used them for that upon return from patrol. Later, when some of our own ships met with trouble in the Atlantic, we left the dummy bombs at home and lugged ash-cans—plain old depth charges which if not dropped properly could be as dangerous to us as to an enemy!—and these we used until the Navy was able to supply regulation depth bombs.

It was not all routine patrolling, however. At times there were interesting sidelights. One day two of our PBYs were ordered to Newfoundland to escort a passenger ship, the *Iroquois*, en route from the United Kingdom to New York. Through our naval attaché in Berlin had come the solemn warning, straight from Germany's Admiral Raeder, that the British were planning to torpedo this American ship in such a way that the Germans would be blamed for the incident. According to Admiral Raeder, the sinking would take place off Newfoundland.

Commander Turner, Lieutenant L. B. "Shiek" Southerland and Lieutenant "Jimmy" Morris took planes out and sat over the ship, just in case. They shepherded her all the way to New York. None of them saw any sign of a submarine, and after a good deal of tension the entire trip fizzled out to nothing, except that Commander Turner had a run-in with a news photographic plane that was hovering too near the *Iroquois* off Ambrose Light Ship. Slightly annoyed by the plane's persistence, the commander ordered his waist machine gunner to point his .50-caliber at the pest and frighten him away. The photographer, in turn, pointed his camera at the waist gunner, and in the papers next day were some beautiful close-up shots of Commander Turner's PBY with the waist gunner grimly aiming his machine gun at the reader!

That was the one and only time we escorted the *Iroquois*, but the

Kungsholm was an old friend of ours as she returned across the Atlantic with passengers fleeing the war in Europe. Escorting her was a pleasant chore, a relief after days of flying over drab, dirty cargo ships and plodding tankers. The boys all bid for it eagerly. Zooming the *Kungsholm*, they could be sure of a gay wave from the pretty girls sunbathing on her deck!

In November, 1940, three of our planes were sent to Bermuda—the first to patrol from that base after its acquisition from the British. Lieutenant Ted Frederick, our flight officer, led the group, and the Bermuda patrol had its full share of exciting moments, especially when the wind blew up to ninety-seven miles an hour and we had to secure our planes to water-filled barrels to hold them down! We hunted many reported German U-boats and searched at times for battleships and cruisers which had broken loose from the North Sea blockade. In spare time we rode bicycles to our hearts' content on that bike-infested island.

For headquarters there we used a PBY wing-box left behind by the British. It was not much, but improvisation was the watchword in those days, and the wing-box served its purpose until it fell apart, at last, from being dragged around the island so much by tractors.

When the President made his trip to the Caribbean in the *Tuscaloosa*, Captain Isbell and I were assigned to carry the presidential mail. We based our two PBYs at San Juan, Puerto Rico. The presidential mail pouch, guarded by a postal inspector, was flown to us there from Washington and we carried it out to the ship whose location of course was kept secret. To be ready in case the President had to be flown back to the mainland in a hurry—an event not at all improbable in those days of international tension—we installed a special bunk in Captain Isbell's plane and constructed a unique rubber-treaded gangway.

Our PBYs winged out to the President's ship almost daily, and on each trip we carried a closely guarded pouch which appeared to be packed with about all it would hold. Though we never indulged in speculation on its contents, I'm sure most of us did some private wondering. With German subs prowling the Atlantic, and the United States girding for conflict, we were certainly carrying something more important than postage stamps!

One morning Captain Isbell and I flew out to the ship together. The *Tuscaloosa* was some 250 miles west of our San Juan base, just south of Hispaniola, heading east. The sea in her vicinity was on a rampage. Great waves paraded beneath us, bursting high in the air and tossing up spouts of spray that looked, from where we sat, like gun-muzzle explosions. Captain Isbell informed the ship that we couldn't land on such a sea without endangering the lives of all aboard.

The *Tuscaloosa* made a slick for us, but even then Isbell was dubious. The slick merely flattened the wave-tops, and there simply wasn't a level area large enough to accommodate a plane. By radio the skipper directed me to stay in the air, and against his better judgment he finally tried it. Easing his PBY down over the slick, he bounced once or twice, straightened out and sat down. But he'd been right. The plane's bottom took a beating and began to leak.

The *Tuscaloosa* lowered a small-boat to pick up the mail and as it struggled through the heavy sea toward the captain a wave caught it, flung it against him, and wrecked the plane's rudder. That finished him.

Helpless to offer assistance, I circled while Isbell's men clambered over the damaged plane to see what could be done with her. She could be taxied, they decided, even though her bottom and rudder were damaged. Isbell radioed me to look for a safe place to anchor. I didn't like leaving him in such a spot but could do nothing there to help him, and obeying orders, at last discovered a small island off the southern tip of Hispaniola directly upwind from him. On the lee side of this island (Beata) I saw what appeared to be a safe anchorage. If the skipper could get there he would at least be out of danger. He began the try, though the distance was seven or eight miles.

The trip was a nightmare for Captain Isbell and his crew. Bucking a wind of gale force, the damaged PBY lumbered over and through waves that gave her a frightful pounding every foot of the way. Isbell had to pilot the ship with his head out the window, because most of the time her hull was submerged by tumbling mountains of water and the rest of the time she was lashed by blinding spray. That he reached the anchorage at all was due purely to expert handling of the plane and a good-sized chunk of fortitude—for if the plane had cracked up under the pounding or taken a dive under one of those sky-scraping waves, we in the air above could have done little to help.

The anchorage, however, was a good one, and when the storm had passed we thoroughly enjoyed our stay there while patching up the plane for the run back to San Juan. We had ourselves a swim and went exploring. I remember discovering in the jungle a little thatch-roofed shack containing an overturned three-legged stool and a board—typically "Robinson Crusoe." But the place must have been used by someone with a political turn of mind—perhaps his political affections had caused him to hide out there for a time— for on the face of the board was burned the inscription, "Viva el Presidente!" One of the boys appropriated it for a souvenir. After taking off, we discovered we had left a souvenir of our own. My third pilot, Aviation Pilot Jones, had forgotten his socks!

Our Neutrality Patrol began to spread its wings as the war clouds thickened. Carrying depth bombs now—for German submarines were more and more frequently violating the neutral zone—we extended the patrol area and operated from a number of new bases. At one time our twelve planes were in eight different places—Argentia (Newfoundland), Newport, Bermuda, Norfolk, Charleston, Key West, San Juan and Trinidad!

For me, however, a change was in the offing. When the *Bowditch* went north to survey the Newfoundland area for naval bases, Captain Isbell and Ted Frederick left us. They flew patrols for the ship and did some mapping in connection with her work. The assignment completed, Captain Isbell returned to be chief of staff to the Commander, Patrol Wings, Atlantic, at Norfolk, and when I landed there on patrol from Bermuda on October 22, 1941, he called me in. That was my last flight in the squadron, and my copilot, incidentally, was a fellow Tar Heel, Ensign "Fuzzy" Fisler, who later became the first man to receive a Navy Cross in World War II.

"Bus," Captain Isbell said, "I've got a job for you if you want it. Something different."

I hesitated. The Neutrality Patrol had been a monotonous chore in the beginning, but lately the assignments had been more and more interesting, and I was not sure that I'd welcome a change. "What sort of job, sir?" I asked.

"You've seen the XPBS around here, haven't you?"

The XPBS was an experimental four-engined seaplane, a Sikorsky job, which had been in and out of Norfolk frequently. Yes, I'd seen her, I said.

"She's yours if you want her," Captain Isbell declared.

I didn't know whether to want her or not. Any day now the old Neutrality Patrol might really develop into hot stuff. One of our destroyers, the *Greer*, had been attacked by a German sub while en route to Iceland in September, and orders had been issued by the President to shoot on sight any vessel attempting to interfere with American shipping. Just a week ago, on October 15, the new destroyer *Kearny* had been torpedoed in the North Atlantic, with eleven of her crew killed and seven wounded. It seemed certain now that our PBYs would soon see real action, and I didn't want to miss it. Besides, I was in line to be made executive officer of VP-51 (alias VP-11, VP-54).

On the other hand, the XPBS might have a future, too. She was a huge plane, the only one of her type in existence, and had been used on some fairly important missions, most of them well off the beaten track. Could be she would go places!

"Well," Captain Isbell insisted, "what do you say?"

"I'll take it, Captain."

"Pack up tonight, then," he told me. "You're leaving tomorrow."

8   *ESCORT FOR MR. CHURCHILL*

IN naval aviation the code letters employed to designate types of
planes are not picked at random; each has its meaning. The more
commonly used codes are as follows:

B, *bomber;* BF, *bomber-fighter;* F, *fighter;* G, *single-engined
transport;* J, *utility;* JR, *utility transport;* N, *training;* O, *observation;* P, *patrol;* PB, *patrol bomber;* R, *multi-engined transport;* S,
scout; SB, *scout bomber;* SO, *scout observation;* T, *torpedo;* TB,
torpedo bomber; X, *experimental.*

Following these designations, the manufacturer's name is indicated by still another letter. SOC, for example, is a scout observation plane manufactured by Curtiss. PBY is a patrol bomber made
by Consolidated—Y being the official designation for Consolidated
aircraft.

In addition, some codes contain numerals. These, too, are of special significance. A PB2Y, for example, is a Consolidated patrol
bomber once improved over the original—the "2" indicates that
this is the second design accepted by the Navy. PB4Y is a patrol
bomber of a further radical design—the fourth.

Finally, the designation sometimes ends with a second numeral,
preceded by a hyphen, as SB2C-1. The final numeral indicates the
model. PBY-4 is the fourth model of the original Consolidated twin-
engined patrol bomber. PB4Y-1 is the first model of the fourth
patrol bomber of different design produced by the same company.
The former is a two-engined flying boat; the latter is a four-engined
land plane.

Much simpler are the popular names given to the planes. SBD
then becomes the Dauntless; SB2C-1, the Helldiver; F6F, the Hell
Cat; PB4Y, the Liberator, et cetera.

The foregoing is written to explain the designation of the four-
engined seaplane which was now my special baby. The XPBS-1
was an experimental patrol bomber, first model, manufactured by

Sikorsky. Insofar as the Navy was concerned, she was still in the testing stage and had not yet been accepted. (She never was.) Because of her experimental status, her engine noses were painted red. Few spare parts existed, and only one spare set of engines.

But she was a big plane—her wing span was 126 feet—capable of flying tremendous distances, and the Navy had found many important uses for her. (When she cracked up at last in San Francisco Harbor she was not replaced and so enjoys the distinction, God rest her soul, of being the only airplane of her kind ever to serve the Navy.)

With Tommy Neblett, the plane's pilot, and John Macroth, co-pilot and navigator, I traveled to Anacostia in the XPBS to take aboard a group of Congressmen for my first mission. They were members of the House Naval Affairs Committee, en route to the Caribbean on an inspection trip. While waiting for them at Anacostia, Tommy and I talked matters over.

"Better use all the spare time you can find," he advised, "to familiarize yourself with the plane. She handles well but has her idiosyncrasies, quite a few of them, and we wouldn't want the Committee to go home with any belittling notions about the Navy. My, no." Grinning, he added sagely, "Better not let them know this is your check-out flight, either. We want them to enjoy themselves!"

We were exceedingly gentle with the Congressmen. A cargo of eggs could not have been handled with more loving care. Not once were they subjected to the tension of a too-long take-off or bumped about in landing.

The trip was strictly a pleasure cruise, even for pilot, co-pilot and crew. With Neblett at the controls and me in the co-pilot's seat, we touched at Bermuda, San Juan and the Virgin Islands, flew over Saint Lucia and the Windward Islands to Trinidad. There, with three or four days to kill and the law-makers gone about their business, I diligently practised take-offs and landings until satisfied that I could handle the plane. After checking out in her—that is, flying her under Neblett's critical eye and qualifying to his satisfaction—I flew her back to Jacksonville by a roundabout route through British Guiana and Kingston, Jamaica. She was not hard to handle and her moods—then—were infrequent. Our distinguished passengers registered no complaints.

Following this trip, the XPBS remained on call day and night for special flights. The calls were not too frequent, however, and though I was satisfied with the job and enjoyed the important-sounding title of "Officer-in-Charge, XPBS-1 Airplane," I was not at all sure that the boys back in VP-51 would not have the last laugh. War was certainly moving closer to American shores. On October 30 the *Salinas*, a tanker, was torpedoed off Newfoundland. A day later in the same vicinity the *Reuben James*, an over-age destroyer, was torpedoed and sunk with great loss of life. On November 17 the Neutrality Act was partially repealed and American merchant ships began to leave port with guns mounted on their decks. This was going to be a shooting war and everyone knew it. But, confound it, the XPBS was not a shooting airplane! We had only two .50-caliber and two .30-caliber hand-held machine guns for armament!

Then on December 7 came the word that Pearl Harbor had been attacked—with rumors, unofficial, that most of our Pacific fleet had been destroyed or damaged. My old PBY squadron, VP-51, was rushed from the Atlantic patrol to Hawaii—the first Atlantic PBY squadron to get to the Pacific after we received "the word"—in a move to strengthen the defenses there against further raids or a possible invasion attempt. The XPBS began to look like a freight car on scheduled runs between Topeka and Tulsa. All of us had long faces.

But the situation quickly changed. One evening soon after the Pearl Harbor attack, I was called to the Norfolk office of Admiral McWhorter, Commander of Patrol Wings, Atlantic, and found there a group of exceedingly important officials, both military and civilian. A sober hush hung over the gathering; no words were wasted in pleasant greetings. Something big was obviously in the air.

The something was Mr. Winston Churchill. He had been in Washington on an official visit and was about to return to London. The utmost secrecy was essential because the intervening Atlantic in those days was swarming with German submarines. None of us,

I'm sure, envied the Prime Minister his journey. For the sub commander who waylaid *that* ship—if Mr. Churchill were planning to travel by ship—the man at Berchtesgaden would probably hand-paint a deluxe medal of honor.

But would the Prime Minister return by ship? He was to be flown to Bermuda, we were told, where a British man-of-war was waiting. Part of our job was to escort him to Bermuda. There we were to stand by for further orders—which might mean anything. The XPBS could easily cross the Atlantic.

I alerted the crew and we waited, trying not to speculate too much on our ultimate destination. A rumor reached me that our part of the journey would *not* terminate at Bermuda, but, sworn to secrecy, I could not even tell Macroth, my co-pilot, where we were going.

At 3 A.M. the Prime Minister arrived with his party by train from Washington. It was a large party; Mr. Churchill was not by any means the only war-important person who stepped off the train. Waiting with us were a British and two Pan American clippers which had quietly slipped into Norfolk a few hours earlier. The Prime Minister and his party boarded the clippers. We, with another four-engined Navy plane, the XPB2Y, were to provide a protective escort.

The trip to Bermuda was quite uneventful. There we waited. Perhaps the rumor that we were to fly Mr. Churchill to England had been circulated with a purpose. At any rate, we were handed no orders. The clippers took off. The warship hauled anchor. Still we waited, only to learn after hours of needless concern that Mr. Churchill had gone on to England in the British clipper.

Had we been held there as a decoy? We didn't know, and don't to this day. Anyway, after we had sat there for what someone considered a sufficiently long time, orders were given me to return to Washington. I took with me the Governor of Bermuda and Her Ladyship, who were on their way to the States for a White House conference.

It was a pleasant trip, with none of the tension we had felt going down. Our passengers, charming people, enjoyed themselves thoroughly, visiting in the cockpit, talking to the crew, asking questions. We let both of them fly the plane for a while and I was amazed at Governor Knollys' smooth handling of it—until I learned later that he had been an RAF pilot in the last war.

Over Washington, however, the weather turned bad and the Anacostia River was closed in. Unable to land, we flew on to Norfolk. And there, awaiting me, were orders such as I had not dared to anticipate. My eyes must have popped as I read them. VP-51 could have its Hawaiian patrol assignment!

"You are instructed," I read, "to take the XPBS-1 Airplane to San Diego by way of Pensacola and Corpus Christi, there to pick up cargo for delivery to the Commander-in-Chief, Asiatic Fleet, at Port Darwin, Soerabaja, or Batavia, via Pearl Harbor." *

It was urgent. The ten members of the crew had time to make a few telephone calls while I hurried home—home was in Norfolk then—to say good-by to Thelma and the youngsters. I'm afraid the farewells were inadequate, for the kids now numbered five: Nancy, Jackie, Sally, Michael and Peter. Peter, a year old, gave Papa a lusty howl from his mother's arms as I waved from the end of the street. They were in for a long wait. Navy wives and children always seem to be waiting.

But, for me, all those years of training had at last led to something. The XPBS and I were going to war.

9   *"THE JAPS WILL USE IT ANYWAY"*

BEFORE the war, the mid-Pacific islands of Wake and Guam were way stations on the clipper route to Manila, the most important trans-Pacific air route then in existence. The Japs put the clippers out of business by seizing the way stations. As for the South Pacific, we had only just started an air route. Travel by air from Hawaii to the Southwest Pacific was in its infancy.

* Official Navy orders are not worded just this way. Commander Miller has "decoded" this one to make it comprehensible to the uninitiated.

The situation in the Pacific looked dark indeed when the XPBS arrived at San Diego. In the Philippines, the Japs had landed on Luzon and were pushing General MacArthur's men back toward Bataan. The naval base at Cavite had been bombed out of existence. At sea, the Asiatic Fleet was fighting to hold the Dutch East Indies, where Admiral Hart had set up his headquarters. But the Japs had too much—and they had air power to support their naval operations, which we didn't. Commander Talbot slowed them with a destroyer torpedo attack in the Battle of Makassar Strait, January 24, but their amphibious forces landed on New Britain and Borneo, and they began reaching for Soerabaja with land-based planes. The outlook for the whole of the Netherlands Indies was gloomy.

At San Diego, repair parts for the handful of Catalinas fighting with Admiral Hart's little fleet were piled up, awaiting delivery. Our job was to deliver them. While this cargo was being added to the load we had taken aboard at Norfolk, we obtained charts of the Pacific and studied them.

The charts were not too encouraging. They told us where to go, of course, but information on the island way stations was conspicuously lacking. Johnny Macroth, my navigator, hunted through all sorts of reference books for additional information while I saw the cargo aboard.

Finding room for all that cargo was a headache, for a small mountain of it had been heaped on the ramp, and the XPBS was, after all, only an airplane. Some of the last-minute priority freight even landed in the cockpit. But at last, with her fuel tanks filled, 7,500 pounds of cargo securely aboard and her engines checked for the umptieth time, old XPBS was ready to go. The afternoon sun lay low on the horizon between us and Pearl Harbor, our initial destination.

Then, just as we were about to cast off, an officer came racing down the ramp, shouting and waving at us. "Wait a minute!" he yelled. "You have to take some torpedo detonators!"

We held the lines and I wriggled out of the crowded cockpit. "Where," I asked him, "are we going to put torpedo detonators?"

"They have to go," he insisted. "Sixty of them. Someone sent the torps out there without them."

The crew patiently rolled up their sleeves again. Obviously, if what was left of Admiral Hart's destroyers and planes were going to keep the Japs out of Java, they couldn't do it with torpedoes that wouldn't explode.

The plane was really loaded then, however, and as we cast off the lines and taxied out for the take-off, my fingers were crossed. We were so deep in the water that the four red-nosed engines had hard work even to move us. In gathering darkness we lumbered over the harbor like a glutted bug struggling to take wing from molasses.

She wouldn't do it. Her nose refused to come up and we couldn't get "on the step"—that is, lift her into the poised, top-of-the-water position from which she would be air-borne. Back we went, well beyond the Coronado ferry crossing for a second attempt. For half an hour we squeezed and shoved the gear farther aft. She behaved a little better with her nose thus lightened, but in gaining speed over the water she threw huge waves up through the props, reducing our power, damaging our props and engines; and the plane was still so logy after the take-off that every time I raised the flaps, we settled toward the sea again. With all that cargo and 3,600 gallons of gas aboard, I was unable, for hours, to gain more than a few hundred feet altitude. She wouldn't go any higher.

It was pitch dark then. The end of day had brought bad weather, and soon the rain was a monotonous, pelting downpour. There were no stars, and as far as we were concerned there was no ocean, though the altimeter said it was just under us. For the navigator this meant trouble. The hop from the West Coast to Hawaii is long—about 2,300 miles—and he needed star sights and drift sights to keep us on course. Our supply of gas was sufficient to get us there only if we didn't waste any in touring the Pacific.

About midnight one of the engines began to give trouble; I lost control of the propeller pitch and had to keep that engine in automatic rich, which meant that we were using considerably more than our calculated output of gas. Still in rain and soup, we were plugging along just over an ocean that we couldn't see, miles from our destination. "We're not going to make it," the co-pilot said. Being skipper, I advised him to "take it easy and don't worry," but knew he was right; we were not going to make it.

That was a long night and an anxious one. Dawn found us still in the soup, unable to get over or under it, and the weather stayed bad all through the morning. The troublesome engine had fouled our carefully formed plans and we knew, 450 miles from Pearl Harbor, that there was not sufficient fuel left in the tanks to take us in. Roberts and Lane, the radiomen, informed Admiral Bellinger, Commander of Patrol Wing Two at Pearl, that we'd have to change course and make for Hilo, which was about 90 miles closer. Then we really began squeezing to keep the old XPBS out of the drink.

Dead tired from flying all night through the rain, I watched the air bubbles dribbling through the flow meter as one tank after another went dry. The Pacific is a big ocean. The wind was strong, the seas below us were high and ugly, and we would probably wreck the plane if forced down. Even if we managed to land safely, we would have to sit for hours awaiting help—and our job was to get those badly needed supplies to the Southwest Pacific in a hurry.

I had a good crew. Nothing disconcerted them much. Lane came into the cockpit, grinning, and waved a paper in front of my face. "Cap'n, we're getting up a pool on when the engines will quit," he said. "Want to place a little bet, do you?"

"When do you think they'll quit?" I asked him.

"Well, we've been in the air nineteen hours and fifty-some minutes. I figure about five minutes more."

Maze, the plane captain, said, "What are you going to do with your winnings? Buy a mermaid?"

I watched the flow meter, waiting for the bubbles from the last tank. Suddenly the co-pilot, who had been looking out the window, grabbed at my arm and said excitedly, "Look! Land!" And it was. It was Hawaii, the "Big Island," on which Hilo was located. The sea was black and sullen, and the whole place had a thoroughly unhospitable appearance, but apparently, despite the balky engine, we were going to get there.

It was too close for comfort, at that. We still had to reach Hilo to find a protected landing place, and XPBS did her best to imitate a snail as we crawled along the coast. As we turned into the wind over the harbor, the plane captain leaned forward between the co-pilot and me and said, "Well, there it is, Cap'n. Bubbles." Our last tank was empty. Air, not gas, was passing through the meter as the plane slid into Hilo and settled sluggishly on the water. There was not enough gas in that last tank to taxi us. Three of the engines wouldn't even cough. The fourth spat for a moment and quit. Unable to reach a buoy about seventy-five yards ahead, we had to anchor in sixty feet of water.

"From now on," Maze said, "the XPBS ought to be called 'Hair-breadth Harry.' Twenty hours and thirty-five minutes in the air, and we land on the last drop of fuel!"

It was about 2 P.M., the sun shining brightly now, but with a heavy ground swell surging in the bay. At frequent intervals during the past hour we had radioed our position to Pearl Harbor. Now we informed the admiral that we would attempt to secure gas and proceed. Marveling at the closeness of our escape, we climbed out on the wing, in the sun, to shake off some of the kinks and dampness acquired in twenty and a half hours of being cramped in the plane with all that cargo. That was when I discovered what had happened to the props.

They were in bad shape, chipped, mushroomed, and saw-edged from the lashing they had taken from the waves during our struggle to get off the water at San Diego. Chunks of metal as large as the tip of my thumb had been punched out of them. The night-long rain had made the damage worse.

Small wonder we had run out of gas! With the props in such condition, it was astounding that we had managed to travel anywhere near that distance before the tanks ran dry!

Repairs were out of the question at Hilo, of course. After sizing up the damage, we tried to signal the people on the beach, a number of whom had gathered by now and were standing at the water's edge, curiously watching us. We waved and shouted, but they would not answer.

Apparently they were suspicious, never having seen a plane like ours before. Our markings confused them, too. Before leaving Norfolk we had painted broad red and white horizontal stripes on the plane's tail, in compliance with a recent order affecting all Navy aircraft; but word of the new markings had not got around yet—certainly it hadn't reached Hilo—and the natives didn't know what to make of us. (We learned later that a Jap submarine, surfacing off the island just a few days before, had shelled the place and demolished a warehouse. So they had reason to be wary.)

But our continued shouting, firing of Very's stars, and energetic waving of the American flag heightened their curiosity, and eventually two men came out in a rowboat. They were still afraid of us and stopped half way, but were close enough to be hailed through a megaphone. I told them who we were, that we needed gas and would have to be towed to a fueling place. With little enthusiasm they at last approached the plane and took two of us ashore, where we hunted up the Standard Oil people to make a deal.

Pearl Harbor, meanwhile, had radioed instructions for us to stand by—a ship was on her way to us with gas. That really made the boys disconsolate. Wait at Hilo for a ship? They might have to wait the whole day and night, when all we wanted was gas enough to get us out of there and on to Pearl, where we might catch an hour or two of sleep. "Tell them we don't want to hang around here, Cap'n," the boys begged. "Let's get going!" So I radioed, and ultimately the situation was squared away with the admiral's blessings. The ship turned about and went back to Pearl, the XPBS borrowed gas from Standard Oil, and we departed.

This was the end of January and the wreckage at Pearl was appalling. Broken ships and parts of ships lay everywhere in the harbor. Great patches of floating oil fouled the water, drifting sluggishly with the tide, clinging to the piers and blackening the shore. Reconstruction was under way, but the hundreds of men struggling to raise sunken ships and repair shattered ones, to clean up the piles of debris and bring order out of the apparently hopeless confusion, were like ants tunneling at a mountain.

My old ship, the *Nevada*, lay among the wreckage not far from the overturned *Oklahoma* and what was left of the *Arizona*. She was no longer the old "Cheer-up Ship," but I felt mighty proud of her, for she had been the only battlewagon able to get under way and try to get out after the start of the attack. I recognized the remains of other once-proud ships of the fleet, too. It seemed incredible that in one assault the Japs could have destroyed so much.

But we had no time for a tour of the naval base. The XPBS, unfit for operation, had to be repaired quickly. She could not be drawn up on the beach; we had no beaching gear. So we taxied her to the Pan American floats at Pearl City and went to work on her there. While the work proceeded, I found time to look up my old Atlantic squadron, patrolling from Ford Island, and borrowed Dick Davis from them to be an additional pilot and navigator for our trip south.

It took four full days to patch up the XPBS. The props had to be removed, honed down, trued, balanced and replaced—an unhealthy operation which reduced their thickness and length and was certain to affect the plane's performance. But without spare parts we had no choice. In those days you did the best you could with what you had. While we worked on the plane, the cargo was rearranged and more added for the shorter hops ahead of us, for now, with less gas to tote, we could fly with a larger pay-load. On the air route to Australia, the longest over-water leg is that from the States to Hawaii. From there on down, the islands of the Pacific are closer together.

On the morning of January 30, ready to go again, we suffered another last-minute delay when the admiral's barge sped out from the pier, signaling us to wait. In the barge sat white-haired, white-mustached General Patrick Hurley, surrounded by his gear and clutching a sack significantly decorated with bank markings. The general had been appointed first American minister to New Zealand and was on his way there, taking with him some $250,000 in "folding money" with which to negotiate for supplies and equipment needed by our forces in the South Pacific. He was to be our passenger as far as Australia. If the XPBS conked out between some of those scattered Pacific way stations, we were certainly going to be missed by the tax-payers!

The take-off was difficult. With the props honed down as they were, we simply hadn't the power to lift such a load of cargo, and I barely got the plane into the air above the booms supporting the anti-submarine nets at the entrance to Pearl Harbor. At Palmyra—not the Syrian city sacked by the Romans, but a pin-point Pacific island about eleven hundred statute miles south of Honolulu—we had another harrowing time. Dick Davis stood in the cockpit and groaned audibly as the plane all but shook herself to pieces in her struggle to be air-borne. "Frankly," Dick said weakly, after we'd made it, "I'd rather be back where I came from."

Palmyra for us was only an over-night pause for refueling, and we were off again for Canton, which lies in the Phoenix group just below the equator. There the war in the Pacific seemed uncomfortably close. With the exception of one aged Britisher and his wife, all the peacetime residents had fled, anticipating a Japanese strike at the area. Pan American Airways had maintained a station there but it too was abandoned, and all Pan Am personnel had left. The sole inhabitants other than the English couple—and she was the only woman on the island—were members of an engineer detachment sent down to build an air strip. They were camped in tunnels and caves, awaiting the time and perhaps the inclination to build more comfortable quarters.

The heat at Canton was intense. Every living soul on the island showed the effects of it. Some, too, were obviously depressed by the war news from the west. "What the hell," one man said, "if we get the air strip built, the Japs will be the ones to use it anyway."

This was a singular attitude, I thought. At Pearl Harbor, despite the destruction and loss of life, the feeling had been one of sullen, angry resolution that the Japs must be made to suffer for their treachery. There was bitterness, but it was aimed at the enemy and was therefore an important weapon of retaliation. Here on Canton the bitterness was vague and pointless; it was no weapon at all but only a form of illness hármful to *us*. The men were living in poor shelters and eating off soap-boxes. They were wearing soiled, sloppy clothes and shoes without laces. They were listless.

One man wore on his arm a ragged, dirty strip of towel held in place with a piece of rope. I asked him what was the matter. "Oh, hell," he said, "some sort of tropical infection."

"Haven't you a doctor in the outfit?"

He looked at me for a moment in silence, and then shrugged. "I'm the doctor," he said. There seemed to be nothing more to say.

On Canton I met a perspiring Army officer who was walking aimlessly around with a shovel and had been digging, he said, for days. I asked what he hoped to exhume. He explained that the Pan Am people, before quitting the island, had carefully buried their meteorological instruments, which he needed for operation of the airfield. "They left me a map," he said, "but, dammit, I'm still digging. Not that I blame Pan American, you understand—if they'd stayed here, the Japs *would* have come calling. But this Captain Kidd stuff is weary work in the hot sun."

There had been a hotel on the island. It was still there, on the side opposite the engineers' encampment. But now it was deserted, except for a lone Army sergeant, and as we walked through the empty lobby our footfalls stirred up strange echoes. The register stood open on the desk, just as the clerk had left it. Keys dangled neatly from the mail slots. Magazines and newspapers lay on the tables. The guests had fled.

"Reminds me," someone said, "of something by Poe. Maybe they all come back once a year at midnight for a ghostly frolic." But the creepy atmosphere did not appear to bother the sergeant. As *maître d'hôtel* he lounged about in shorts, sloppy shoes with no laces, a little fore-and-aft cap with a feather stuck in the side, and seemed to be thoroughly enjoying the informality of his job. (A little while before, as "beachmaster," the same fellow had brought our plane to the dock for gassing. Apparently he had many "official" duties!)

Prowling about, we discovered some tennis racquets and balls behind the desk, and with these we amused ourselves on the hotel court, free of charge, until the heat put an end to such foolishness.

If the proprietors should happen to run across this account, they'll probably send me a bill.

On the other hand, perhaps they won't. For when we stopped at Canton on the way back from Australia, this abandoned hotel had been taken over by 1,500 Army men who had stripped it, transformed it into a barracks and moved in, bag and baggage. There were no unlaced shoes on Canton then. The atmosphere had changed. Antiaircraft units were unloading their weapons with a spirit that plainly dared the Japs to call around any time and meet a determined reception committee.

10   *SO LITTLE, SO LATE*

By now we were a tight little crowd aboard the XPBS. Dick Davis spent most of his time in the cockpit, telling us of his adventures in the Pacific. General Hurley, when not visiting up forward, could be found back among the cargo, playing cribbage with the crew.

The general won most of the cribbage games. Possibly the huge six-shooter he carried had something to do with his invulnerability. He is a Westerner, and with his white hair and fine white mustache *and* the six-shooter, he bore a striking resemblance, so the boys insisted, to Buffalo Bill Cody. The weapon, he solemnly informed us, was for protection. "A quarter of a million dollars is something I don't tote around every day!"

En route to our next port of call, Suva, we crossed the International Date Line. Approaching the line I was celebrating my thirty-fourth birthday. A moment later the date became February 2—"before we could even bake you a cake, Cap'n!"—and the celebration was over. But, though we didn't know it at the time, there were bigger things to celebrate that day. My old Pensacola commandant, Admiral Halsey, had steamed out from Pearl Harbor with a force of carriers, cruisers and destroyers, and on February 2 he was raiding Japanese-occupied islands not far to the west of us. With him was Rear Admiral Frank J. Fletcher, USN, and the blow they struck was the first against the Marshalls and the Gilberts. Months later I was to know some of those islands, particularly Kwajalein, Wotje, Tarawa and Apamama, rather well.

But the Japs were busy that February 2, too. In the Dutch East Indies, for which we were bound, they were seizing new positions on the coast of Borneo and dealing out murderous punishment to the PBYs of PatWing Ten, which, with the little Dutch fighter force and the remnants of our Flying Fortress and P-40 squadrons pushed down from the Philippines, were all the "air support" on which Admiral Hart could call in attempting to hold back the enemy advance. (The PBY in which I had patrolled the Atlantic, incidentally, was part of this battered little air force, for my old Neutrality Patrol squadron, arriving at Pearl, had turned its PBY-5s over to VP-22 and was patrolling out of Pearl in obsolete PBY-3s. VP-22, already familiar with the Central and South Pacific, had taken the more modern planes to the Indies.)

Still struggling under a good deal more cargo than Mr. Sikorsky had ever intended her to carry, the XPBS at last arrived at Suva, the capital of Viti Levu in the Fijis. Suva used to be the romantic center of the so-called South Seas. (Whatever happened to that phrase "South Seas"? All you ever hear now is "South, Southwest, or Central Pacific.")

Suva used to be famous, too, for its Grand Pacific Hotel, which played host to celebrities from all over the world. And for other things, as well, which are better left unmentioned. It hadn't changed much in February, 1942. The hotel was still crowded, especially the bar; the servants were still barefoot Indian natives (the Fijis are British); the wild-looking, mop-haired Fiji Islanders were still astonishingly soft-mannered and gentle. I had a drink at the bar with Dick Davis, the first since San Diego, and then, tired after the long grind of 1,200-odd miles from Canton, we went to our room.

Dick eyed the bed and sighed his appreciation. It was a beautiful bed. He said he thought Sherman was wrong about war and luxuriously stretched himself out, without pausing even to shake off his shoes. Then he began squirming. With a puzzled look he sat up and drew back the sheets. "They put rocks or something in the beds out here," he grumbled. "Must be a British custom. Damned uncivilized."

He found what he was groping for and hauled it out. It was not a rock. It was an unopened fifth of the best Scotch whisky, which someone must have hidden there and neglected to remove!

Our take-off from Suva was a nightmare. Lacking both wind and room, we twice thundered the length of the abbreviated harbor without gathering speed enough to rise from the water, and twice narrowly missed having accidents. I was getting used to this sort of thing by now, but still I sweated mentally and physically through every moment of the take-off runs. Nor did I sweat alone. There was a moment of decision on these take-offs beyond which a man could not change his mind; a moment when, having attained a certain amount of momentum and used up a certain proportion of the available space, the plane had either to take wing or hit something. It was a ticklish business and gave all of us a laundry problem with shirts that wouldn't stay dry.

After futile attempts to get the plane up "on the step," I gave up and requested the Pan Am boat there to lead us through the tortuous passage to Luathala Harbor, where the take-off run was longer. We cleared Luathala on the fifth attempt, all of us exhausted from the strain. Then, after misbehaving so badly, the XPBS flew like a placid old cow to Noumea, New Caledonia, without even a wiggle of protest.

Noumea was not then the bustling naval base from which our forces later swept northward to drive the Japs out of the Solomons. The situation then was critical, and the senior United States officer on the entire island was a young Army lieutenant named Sauer who had been sent with a small force of engineers to build some airfields. He was waiting for us when we stepped ashore, and the relief that lightened his dour face when he saw General Hurley would have been comical had not the reason for it been so serious.

"Sir, we're in a ticklish spot here," he said. "If the Japs wanted to, they could move in any time. We haven't a thing to stop them with."

The general immediately went with Lieutenant Sauer to visit the governor of the island. Later he called at the hotel and asked if I cared to go for a little ride. He was obviously troubled and wanted someone to talk to—not for advice, but for an audience. (General Hurley is an old hand at trouble-shooting, and at this writing is United States Ambassador to China.)

As we rode through the streets of Noumea, past the shops and neat little French homes with their colorful flower gardens, the general told me how critical the situation really was. The whole island, he said, was garrisoned with only a handful of New Zealand troops armed with a few rifles and a couple of machine guns. Air defense was nonexistent and the island had no heavy weapons at all with which to repel an invasion. The Japs could take the place without working up an appetite.

The Japs probably knew the situation on New Caledonia as well as we did, but they were busy at the moment to the westward, trying to remove Admiral Hart's little fleet of American, Dutch and British warships from the Dutch Indies. It was not much of a fleet. The American cruisers *Houston*, *Marblehead* and *Boise*, with the British *Exeter*, the Dutch *Java* and *De Ruyter* and two Australian cruisers, were its principal components. Its air support, never strong, had been whittled down to nothing. With this defensive force wiped out, the Japs would pluck not only New Caledonia but other, fatter prizes as well. We would have nothing to stop them.

Yet in Noumea the people were complacently strolling the streets while their government, annoyed by our hesitation to recognize the Free French, was protesting the efforts of Lieutenant Sauer and his engineers to build airfields. No wonder white-haired General Hurley was worried. "They don't realize their danger," he told me, shaking his head. "They just don't see it. The Japs could have this island tomorrow."

The harbor at Noumea is long and narrow, shaped something like a fish. We began our take-off with the XPBS well back on the fish's tail, and for five minutes and forty-five seconds by the co-pilot's watch, the overloaded plane roared across the harbor at seven inches above allowable manifold pressure, refusing to climb. Then, when we thought the four laboring engines would leap right out at us, she tipped her stubborn nose up. It was a close call. "One or two more like that," said the co-pilot, "and we'd better wire ahead to reserve some padded cells."

We flew west across the Coral Sea to Australia, through increasingly bad weather which turned into a driving rainstorm at Townsville. Despite the rain, we managed to squeeze in and make a landing. It was either then or on the way back—the time is not important—that a man came out to us in a boat and solemnly advised us to keep out of the water. There was a slaughterhouse near by, he said, and sharks were attracted by the refuse. "Very many sharks. Hundreds of them. One time a man bet he could swim the channel here, and he did, but on the way back—pouff!—the sharks got him. So you stay out of the water please, yes?" We stayed out of the water.

From Townsville to the next way station, Groote Eylandt, the route lay over Australia and we had a ridge of mountains to cross. The XPBS had no love for mountains—not with all that cargo and half a dozen hitch-hikers aboard. Straining for altitude, she all but rubbed her belly on the way over the humps. The weather, meanwhile, had apparently decided to make the last few miles of our trip as difficult as possible, and it worsened as we went along. The hitch-hikers were noticeably uneasy.

Groote Eylandt is a tiny island on the northwest side of the Gulf of Carpentaria, which bites a large, deep segment out of northern Australia, and when we dropped out of the sky there we found a few million voracious mosquitoes and nineteen men. None of the men had been off the place in two years.

With the Japs on the rampage in the Indies, and nothing to stop them from pouring down from the islands to the north to seize this isolated little outpost, our hosts were alarmed and anxious. "When is help coming?" they asked with a kind of forlorn resignation. "When is help coming?" What could we tell them?

Because the mosquitoes were eating us on the hoof, we left as quickly as possible and made for Darwin, where I hoped to find Admiral Glassford. My orders were to report to him at Darwin "or wherever he may be." He was not there, but the *Langley* was, and Commander McConnell, her skipper, said that the admiral was at Soerabaja. After we had fueled astern of the *Langley* and moved to a buoy, the commander sent a dispatch to Soerabaja to say that I was coming. But the XPBS needed a change of spark plugs and an engine check before venturing into an area so jammed with Japs, and so for two nights I bunked on the ship. (The *Langley*, our first carrier, had been converted to a seaplane tender and was one of four such tenders serving the PBYs in the area. She was waiting at Darwin for an expected cargo of fighter planes en route from the States. Later in the month she was sunk off Tjilatjap while attempting vainly to rush these fighters to the aid of our battered forces to the north.)

While the plane was being repaired, I visited the town, and heard everywhere the same old question: "When is help coming?" But now there was genuine desperation in the asking. The war was very close to Darwin. The town itself was like something out of a Wild West movie, its unpaved streets jammed with army vehicles, its frame buildings packed with people anxiously awaiting transportation by rail and Defense Highway across the great Australian desert to the less vulnerable cities of the south and east.

These people were bewildered, and some were bitter. Many had what can be described only as a wild look in their eyes. The tremendous momentum of the Japanese advance, the seizure of the northern Indies and the impending fall of Singapore seemed to them incredible.

Well, it seemed so to us, too, for we had come into the war zone so quickly that the news of the campaign had not caught up with us. Commander McConnell on the *Langley* had told me something of the gravity of the situation, but my greatest shock was in meeting what was left of the West and Southwest Pacific Patrol outfits. They had been flying their lumbering old PBYs in the islands and knew what it was like there. With other PBY men in the area they

called themselves the "Cold Turkey Patrol," so slight was their
chance of survival in combat with the Jap fighter planes which
daily grew more numerous throughout the Indies.

I knew these boys and knew their capabilities. Yet of the twelve
planes of my old outfit which VP-22 had brought to Australia and
Java, only two were then in action. One was my old plane, No.
2308. The others had been shot down or destroyed on the water,
and many familiar faces were missing among the pilots. When I
said that I was bound for Soerabaja with repair parts for the PBYs
there, they advised me soberly not to waste the time. "You won't
find anything to repair when you get there," they declared.

With Commander McConnell, I discussed the flight to Soerabaja
and the prospects of getting there without mishap. Together we
laid out a route. I would fly due west from Darwin to 120° E.
Longitude, and then head for the Bali Straits. The Japs were hitting
Soerabaja from land bases on Borneo, from bases to the northeast,
and from carriers roving the straits west, north, and northeast of
Java. My approach would be as nearly the exact opposite of theirs
as was possible.

"One more question," I said when we had finished. "What about
my torpedo detonators? My orders are to deliver them to the senior
naval officer in the area, presumably at Melbourne or Sydney."

"They don't need your detonators there," Commander McConnell replied gravely. "The fleet's in the Java Sea."

"But I don't want to risk flying them to Soerabaja unless we can
be sure," I protested. "If I'm shot down, they'll be lost."

The commander agreed that there was a fairly good chance of
my being shot down. The Japs, he said, were spreading so swiftly
through the Barrier Islands—Sumatra, Java, Bali, Timor and Cel-
ebes—that no one knew for certain where they might pop up next.
Still, he thought the detonators would be needed at Soerabaja and
that I should take them along. He sent a message, asking for clarifi-
cation of the matter, but at midnight there had been no answer
and so I turned in.

At 3 A.M. a messenger from the Communications Office shook me
awake. Word had come from Java urgently advising me to change

my proposed route; 120° E. Longitude was dangerously far east. I was advised to proceed west to 115° E. Longitude before cutting north. In short, the Jap fighters and bombers were ranging east of Soerabaja along the Barrier Islands, and I would have to extend my westward reach beyond them before turning north. Moreover, I would have to hurry if I hoped to get there at all, for reports were coming in of a naval-air action in which the Japs had hit our little fleet a paralyzing blow. A striking force of cruisers and destroyers under Dutch Admiral Doorman, attempting to close with the enemy at Balikpapan (Borneo), had been attacked by land-based bombers which had put the *Houston* and the *Marblehead* out of action with serious damage. Our grip on Java, my destination, was rapidly weakening.

11   *COLD TURKEYS*

Above the Timor Sea, the sky was layered with shades of gray, dotted here and there with great island-like masses of clouds. The XPBS used these islands of the sky for protection. Through them we warily felt our way westward against strong head winds, alert for enemy fighters as we stair-stepped upward toward the Bali Straits. Over that narrow strip of water hovered a massive black thunderstorm, hiding the precipitous mountain peaks on either side. We barely slipped through.

Our plan was to arrive at Soerabaja just before dusk, after the Japs had expended their daily quota of bombs and retired. It worked without a hitch, and just as the rim of the sun slid under the western peaks of Java, we made a landing in the harbor.

I had been told that Soerabaja was a pretty place, its buildings white and modern, its broad, clean downtown streets as attractive as those in any of our swank southern resorts. "You blink in the bright tropical sunlight and think you have wandered, by mistake, down Miami's Lincoln Road." The naval base, I'd been told, covered an orderly two miles of waterfront.

No doubt this had been true once. But in February, 1942, the city was momentarily expecting invasion and the harbor was not orderly at all. Everywhere lay bomb-wrecked PBYs and the burned-out skeletons of Dutch Dorniers. A typhoon could have done no more damage.

The Japs had been hitting Soerabaja hard. Two of their bombs had exploded on a hangar containing a PBY. Another bomb had leveled the Bachelor Officers' Quarters and forced the men living there to move out with what remained of their possessions. The day I arrived, another bomb had hit a street car packed with people.

The Dutch, quite properly called "phlegmatic," were calm enough, even with the certain knowledge that the enemy advance would soon sweep over them. Mildly bitter, they were looking to their defenses and planning on giving the Japs a savage welcome. Obviously, though, any welcome they could arrange would be pathetically brief, for they stood quite alone; the natives, terrified, had fled from the city to the hills, and it was impossible anywhere in the city to find workers for the thousand and one jobs that needed doing if the defense against invasion were to be anything more than a mere show of courage.

Bitterest of all were our own aviators, pilots and crewmen of the PBY squadrons to whom I had brought replacement parts from San Diego. They greeted me with shrugs. "The stuff is no good now," they said. "Throw it away."

Few of these men expected ever to get home. Some at the air station were auctioning off their belongings—watches, rings, anything they owned that was worth a dollar or two—so that they might send money home to wives and families while there was yet time. They were tired, very tired. And angry.

Not enough—not nearly enough—has been written about the heroism of these men of the "Cold Turkey Patrol" who flew their old PBYs against the Japs. They failed to stop the enemy's advance, it is true, but what we learned from them was later of enormous value in other combat areas, and the dogged courage they showed in the face of really frightful odds should not be forgotten.

Don't misunderstand me, please. The "Cold Turkey Patrol"

was not made up of glamour boys wearing their patriotism in haloes. They asked no eulogies. Many of them were morose and bitter. But they had done, and were still doing, a tremendous job.

Most of them had fought in the Philippines, whence island by island they had been forced southward through the Indies, leaving behind wrecked planes and dead comrades, along as bloody and discouraging a trail of retreat as American fighting men have ever known. At Soerabaja and Darwin were men who had been shot down at sea and had lived through sagas of life-raft suffering as terrible as any yet written. They were still going out daily to face the same victory-flushed Jap fighters. Others told harrowing tales of walking back from parachute landings on stinking, malarial jungle islands. They, too, were still fighting. In lumbering flying coffins that had neither the maneuverability nor the fire-power to give them a chance, they were trying hopelessly to provide "air protection" for our fleet against the best pilots and the best planes in the Japanese air force.

"When you get back," they told me, "tell them to send us some fast, heavily armed land planes. No one can stop the Japs in these boxcars. Get us some B-24s or some B-17Es. And get them in a hurry!"

The B-17E, they said, was the most effective airplane in use in the area, but only a few of these Flying Fortresses which had fought their way down from the Philippines were still operational, and the supply of spare parts had been almost exhausted. The men who flew the Fortresses were chalking up an amazing record, however. The planes were well armed—they carried ten .50-caliber machine guns and one of .30 caliber—and their performance at high altitudes was good. One group, attacked by a squadron of Jap Zero fighters, had shot down at least fifteen of the enemy in a running battle over the Barrier Islands.

As for the B-24s, they too were good against the enemy's fighters, but only if equipped with turbo superchargers. I was told of a group of B-17Es and B-24s which had been attacked by Zeros while in company on a bombing mission. "We climbed for all we were worth and got away from them," said a Fortress pilot, "but the B-24s, with no superchargers, fell behind and were cut out. Several were lost when the Japs concentrated on them. You need a first-rate airplane to hold your own against those Zeros."

That was the word everywhere: the Jap Zero, with its retractable landing gear and large radial air-cooled engine, was an excellent plane, not particularly fast but astoundingly maneuverable. A surprising feature was its long range. Soerabaja was being hit daily by bombers *and* fighters based in Borneo, some four hundred miles distant, yet the fighters often stayed over the target for two hours or more. They carried a 77-gallon fiber belly tank which was jettisoned when empty, and they were armed with four light machine guns and two 20-mm cannon.

"The monkeys are smart with their guns, too," a PBY pilot told me grimly. "You can forget the old theory that the best PBY defense against a fighter is to get low over the water and stay there. Against the Zeros it doesn't work. The Japs simply walk their small-caliber splashes up to the PBY's tail and then open up with their cannon. In other words, they use the light guns for spotting on the cannons. It's murder."

"What defense," I asked him, "do you use?"

"Anything. Anything at all. One trick the boys have worked out is to keep a man in the bow turret facing aft, to tell the pilot, via interphone, what evasive maneuvers to take in order to avoid the fire of fighters on his tail. That helps—sometimes. Hiding in the clouds helps, too, when there are clouds. We lie awake nights dreaming up trick ways of staying alive. Most of them don't work."

I asked many questions, because I had been instructed to. The answers, however, were there at Soerabaja for anyone to see. The situation was hopeless. The enemy then had been hitting the harbor and the city for six or seven consecutive days, arriving between eight and eight-thirty in the morning, attacking in relays all day, and departing for home about two or three in the afternoon. His formations often contained from fifty to eighty planes. Not one of those formations had yet been broken up by our handful of defenders.

The Japs had hundreds of planes to throw into the assault. For defense we had a few Flying Fortresses and B-24s, eight PBYs, six of which had been given to us by the Dutch who called them "death traps" and "suicide boats" and obviously were not at all eager to fly them, and twenty P-40 fighters based sixty miles south at Malang. The P-40s were the only planes we had that were on a par with the Jap Zeros, and even they were none too impressive. I talked to an experienced lieutenant who had witnessed a dogfight between three of them and a Zero, and this is what he told me:

"The fight took place less than four hundred feet over the water, and the Jap was good, very good. He knew his plane and what it would do—and it would do plenty. Out-performing the P-40s, he shot one of them down. Then his guns jammed, or he gave out of ammunition, and he took to his heels with the other two P-40s after him. Our fighters said later they had shot him down at sea, so obviously they were fast enough to overhaul him. But over the harbor, where maneuverability counted for more than speed, the Jap more than held his own."

The same lieutenant had witnessed most of the raids on Soerabaja. "The bombers come in very high, from 24,000 to 27,000 feet, in perfect formation," he said. "Our antiquated AA batteries can't touch them at that height and they know it. They take all the time in the world on their bomb runs and do enormous damage. Their fighters, coming down to strafe, fly very low. Lots of times I could have knocked some of them down by tossing slugs in their path, if I'd had a bucket of nuts and bolts and been in position. You can see for yourself—" and he pointed to the wrecked planes that lay in the water at the air station—"the kind of damage they do."

This, then, was what the men of the "Cold Turkey Patrol" were up against, and, I repeat, their story should be better known. Perhaps the grimmest record of all was that of Patrol Squadron 22, which in a month of operation had lost ten of its twelve PBYs. Five had been shot down in the air; five had been destroyed on the water by strafing or bombing. On one bombing mission in which six of this squadron's planes had taken part, four had failed to return.

"Send us," these men begged, "something to fight with! Send it soon. Tell them if we had a couple of hundred fast, well-armed land planes here and maybe fifty submarines, we could stop the Japs or come mighty close to doing it. Tell them that!"

All this was depressing news to the crew of the XPBS. Obviously our trip to Java had been a waste of time, and the PBY parts which we had brought all the way from San Diego would do no one much good. There still remained the torpedo detonators, however, and so, without much hope, I continued my search for Admiral Glassford. His headquarters, I learned, had been set up in a resort hotel back up in the mountains. But a hotel in the town housed a sub-headquarters, and after stumbling around bomb craters in the blackout, I eventually found the place.

There something good finally came of it all, for when I met the admiral's duty officer and introduced myself as skipper of the XPBS, he greeted me with enthusiasm. "Have you got our torpedo detonators?" he asked eagerly.

"Thank God," I said, "I've at last found out where they're going!"

His own relief overshadowed mine. "Miller," he said soberly, "we really need those detonators. They represent just about our last chance of putting on a decent show here. Get them to me as soon as you can."

The plane was even then being unloaded. At least, I thought it was. With great difficulty we had managed to obtain some natives to get the cargo out and gas the plane for the return trip to Australia. But when I returned to the harbor I learned that the natives had failed to show up. We had to do our own unloading and refueling, after all.

By this time we had been at Soerabaja ten hours or more, and dawn was not far distant. I had been advised by the people at the air station to leave that night and not wait for daylight, but had decided against it. The harbor was so cluttered with wreckage that even taxiing through the channel in the dark, for the take-off, would have called for some extremely delicate piloting. Furthermore, the tide was low by the time we had finished gassing the plane, and there was danger of being caught in the mud of the narrow channel, a prospect even less inviting than that of waiting for daylight and the probable arrival of the Japs.

In the morning we left, and even in daylight it proved to be a hazardous job. A stiff breeze whipped the harbor. I had to use a lot of port motor to keep us from swinging around, and in the process nearly ran down the man who was leading us in a small motorboat. His face, when our thundering red-nosed engines loomed over him, was white with terror, as it had a right to be. He leaped to his feet to dive overboard, and by God's grace I saw him in time.

It was an ugly day. Warned of a violent storm raging in the straits, I had obtained permission from the Dutch authorities to fly straight over Java through a gap in the mountains—where, incidentally, there might be less risk of running into the ubiquitous Japs. But the winds over the mountains tossed us about like a kid's kite. Most of the time the sky was a swollen black sponge, shot through with rain and lightning, squeezed down upon equally black peaks through which our lane of safety was uncomfortably narrow.

The Japs were wise enough to stay out of such weather. They hit Soerabaja that morning from the direction in which I had come the preceding day. But though we heard the air raid over the radio soon after leaving the harbor, we ourselves avoided a challenge by flying straight south before turning east, and with strong west winds and a lightened load we made the 1,200-mile trip back to Darwin in a little over eight hours. (The outbound flight had required thirteen hours.)

From Darwin to Townsville, the XPBS began to protest the punishment she had absorbed. Over the Australian mountain ridge, one prop began running away and had to be feathered. If this had happened on our westward flight it probably would have wrecked us, but now with the load lightened, the plane limped over the peaks on three engines. Despite rain and thunderstorms, we managed to get in at Townsville and there located the trouble. The armature of the prop motor on No. 4 engine had sheared off.

We had no spare prop motors, nor, of course, was there a plane of the XPBS type at Townsville from which one could be pilfered. The crew, however, was used to this sort of thing. They had been hand-picked for the job with just such emergencies in mind, and were seldom stumped. The prop motor they snitched from a P-40 was the wrong size and voltage, but they ingeniously shimmed it up and made it work. After a hair-raising open sea take-off in rough weather, away we went to Noumea.

General Hurley had done something about the situation in Noumea since our stop there on the way out. Soon after we landed in the harbor, a launch came alongside the plane and a guttural voice called up, "Is the captain aboard?" I asked what was wanted. "Want to see the captain," the voice insisted. Coming aboard, the speaker introduced himself as skipper of the *Southern Seas*, the former Curtis yacht which had been sold to Pan American as a floating hotel. Obviously something of importance was on the skipper's mind. He insisted on my returning with him to the yacht.

I finally agreed. But because it was dark, and you never knew with whom you were dealing in those days, I took the precaution of strapping on a gun. It turned out to be a needless precaution. Aboard the *Southern Seas* was Lieutenant Colonel Rich, who had been rushed down to meet the need for more military authority in the area. With him was a St. Louis engineer, Jack Sverdrup, in charge of constructing airfields on a number of South Pacific islands.

The luxury of the *Southern Seas* was a far cry from the cramped quarters of the old XPBS, and I remained aboard over night, reveling in comforts previously enjoyed by Pan Am's customers. But the yacht was of no further use to Pan Am. Their service to New Zealand had been discontinued. The yacht was to be turned over to Sverdrup to serve him as a floating office which could follow him around among the islands. Sverdrup himself, full of war-important plans for the establishment of new air routes (in case the Japs seized some of our island bases), was eager to return to Pearl Harbor. The question: could he travel in XPBS-1 and thereby save precious time?

I had no authorization to take him, of course, but it was impossible in those days to wait for formal instructions. Men were in a hurry and the situation called for on-the-spot decisions. When the XPBS left Noumea the following morning, Sverdrup was aboard. Possibly he regretted the decision later, for the plane by this time had really become cranky. Her engines faltered on the slightest provocation, often with no provocation at all. We were lucky to reach Suva without mishap.

Suva was Sverdrup's headquarters and the scene of his most ambitious project, an airfield on Viti Levu. It was almost the scene, too, of the XPBS's last gasp. On the take-off, sensing something wrong with No. 4 engine, I cut the throttle and discovered that the prop motor had fallen off into the water. So, for the first time on the trip, the crew had an opportunity to "unwind" for a day or two.

We needed this unwinding. In an experimental airplane without spare parts we had traveled thousands of miles over long stretches of the Pacific, landing in tight little harbors where a plane of that size had never before ventured, and, what is worse, taking off under the same acute handicaps. We had done our share of sweating.

Our P-40 prop motor was irretrievably gone, and now we somehow had to repair the one which had gone bad over Australia. This the Department of Public Works at Suva undertook to patch up for us, and so, with nothing to do, we sat around waiting. The *Southern Seas*, having followed us from Noumea, came in while we waited. Aboard her, and at Jack Sverdrup's home on the knoll overlooking the harbor, we were treated as handsomely as visiting celebrities.

From then on, the XPBS suffered from chronic engine trouble and our progress was part limp and part prayer. But she was a good plane; despite her ailments she at last delivered us safely to Pearl Harbor. There I learned that Sverdrup had asked that I be assigned to survey the new air routes on which he was working, but the plane was too far gone for such an undertaking. I flew her to San Francisco and then to San Diego, where she was overhauled and turned over to the Naval Air Transport Service. On her next trip, with Admiral Chester W. Nimitz aboard, she came to grief on landing in San Francisco harbor and turned turtle, killing a co-pilot. That was her finish.

Good old XPBS-1. She had been a dependable plane for many months, and I was sorry that on her one great mission she had been able to accomplish so little.

[*Commander Miller, a modest man, is here being a bit too modest. This trail-blazing flight to Soerabaja in the XPBS earned for him the Commendation Ribbon, and a better estimate of the value of the flight is to be found in the words of Admiral P. N. L. Bellinger, Commander of Patrol Wing Two. Writes Admiral Bellinger: "The completion of this wartime mission, so successfully accomplished under the most exacting conditions imposed by radio silence . . . the conduct of the flight, the leadership, initiative and fine judgment displayed by Lieutenant Miller in preparing his plans and executing them in such an exemplary manner . . . are worthy of special commendation. Commander Patrol Wing Two congratulates him and the officers and men of his plane crew on this fine performance."—H.B.C.*]

12   *THE THUNDER MUG*

THE XPBS had been one of the first couriers from the Southwest Pacific combat area, and on my return from Soerabaja I reported at once to Admiral Nimitz at Pearl Harbor. The admiral led me to a map of the South Pacific, thrust a pointer into my hand and said, "Start talking. Tell us what's going on out there."

I told him, and later told Admiral McCain in San Diego, what I had learned of the performance of our PBYs, B-17s and B-24s in combat with the enemy, and what I had seen of the situation in the South Pacific. Then I wrote out a formal report. Perhaps this report influenced the Navy's decision to acquire some multi-engined land planes; perhaps it didn't. At any rate, in March I was ordered to begin patrolling with the Army in B-24s and Lockheed Hudsons. My job was to learn all I could of the operation of these planes, so that I would be in a position to instruct others.

Then and there I was sold on the Consolidated B-24 Liberator as the airplane most needed by the Navy for long-range patrolling. The Liberator was big enough to carry a really destructive bomb load, powerful enough to cover the long distances of the Pacific, and well enough armed to give a good account of itself against enemy opposition. It was a high-altitude plane with tremendous low-altitude potentialities. I argued volubly that the Liberator would be worth its weight in gold when we began to get back some of the islands which the Japs had overrun. Then, impatiently, I waited for the Navy to acquire some of these airplanes, so that I might go to war in one.

The Navy did acquire some, but not for me. Instead, my old Atlantic Patrol squadron took them over and, reformed as Bombing Squadron 101, became the first Navy unit to operate with the new planes. As for me, I continued to do some patrolling with the Army, and occasionally flew Liberators to the East Coast for the Ferry Command.

Then came the big disappointment. Instead of being assigned to a Liberator squadron, I was ordered to the Naval Air Operational Training Command at Jacksonville and assigned to the staff there. Someone apparently thought that men of thirty-four were past the peak of combat flying. My job as navigational training officer carried with it a raise in rank, upping me to lieutenant commander, but when I picked up Thelma and the kids in Norfolk and traveled with them to Jacksonville to find a home, I was miserable. God knows the Navy needed navigators, but after my trip to Soerabaja I was convinced that the Navy needed combat pilots, too—especially pilots who had flown multi-engined land planes.

I needn't have worried. The job at Jacksonville was important, too. There, in addition to navigation, my pigeons were air-crew and ground-crew training and free gunnery instruction. Again I learned as much as I taught, for the boys who came to Jacksonville had already been through technical training centers and were up for advanced work at operational school. That is, a man reporting as a gunner already knew gunnery; he had only to be taught how to operate the guns aboard an airplane and to function as a member of the plane's crew. In a year of this work, I learned from the men themselves much that was enormously useful when I was given my own squadron.

The good word came in July, 1943, helped along by my good friend and boss, Commander "Skee" Erdmann. "Lieutenant Commander Norman Mickey Miller is hereby detached (from Jacksonville) and will report for duty at San Diego, California, involving flying as Commander of Bombing Squadron 109 when commissioned."

We hadn't much time for training, but the men of VB-109 were exceptional in their ability to learn quickly. Many were veterans. Some had been through tours of duty in the Aleutians, and one, Lieutenant (jg) Amos Shreiner, had even flown with the RAF and had twenty-one operational flights over Europe. Probably some of these veterans were annoyed at having to go through the fundamentals all over again, but they realized the need for it. The squadron had to learn to operate as a unit, no matter how complete the education of its individual members. And so our training was intense, both in the pre-commissioning period and later. Emphasis was upon familiarization, getting to know one another's capabilities, and upon instrument flying, navigation, night flights, gunnery and bombing.

On November 4, the squadron moved west across the Pacific to the Naval Air Station at Kaneohe, in the Hawaiian Islands, and was transferred to Fleet Air Wing Two. There, while the training continued and some of the planes took part in patrols, all of us kept an anxious ear tuned to reports coming in from the South and Central Pacific.

In the South Pacific, the Solomons campaign was about over, with American troops in possession of all the Solomon Islands except Bougainville, where fighting was in progress. In the New

Guinea campaign, Allied troops had rolled the Japs back as far as Finschhafen, on the eastern end of the Huon Peninsula. In the Central Pacific, of particular interest to us, American troops had landed in the Gilberts (November 20), capturing Makin, Tarawa, and other islands of the group, while carrier task forces invading the Marshalls had launched heavy attacks on Kwajalein and Wotje (December 4). It was about time for our number to be called.

The boys of my own plane crew, meanwhile, had been industriously seeking a name for our plane—a practise frowned upon in high circles but permitted without official sanction in combat areas. Dozens of wild and wonderful suggestions had been propounded, and I had thought the matter settled on the flight from San Diego to Hawaii, when the crew had decided on the name *Hit Parade*. Such a name, they insisted, had elastic possibilities. "We can call the top turret *My Blue Heaven*, Cap'n, and the bow turret *Lead Kindly Light*, and so forth." But to their intense disappointment they found an Army Liberator whose crew had beaten them to it, and so once again the solemn discussion of a name got under way.

It would be nice to claim that I selected the name *Thunder Mug* myself, using my wife's initials (*Thunder Mug*, Thelma Miller—Get it?) and I've been accused of that, but it just isn't so. The name originated in the fertile imagination of Elmer Kasperson, my co-pilot. I've been accused, too, of grabbing plane No. '08 for my own because I was born in '08. That isn't so, either. It was pure coincidence. However, '08 became the *Thunder Mug*, and after the boys had painted a huge chamber pot on her sides and ingeniously changed the numerals '08 into "Ye Olde 8-Ball," everyone seemed satisfied. Other planes in the squadron presently became *Mission Belle, Flying Circus, No Foolin', Sky Cow, The Stork, Our Baby, Come Get It, Helldorado, Climbaboard, Urge Me, Consolidated's Mistake, Available Jones, Mug Wump,* and *Sugar Queen*. A few of them, notably *Urge Me, Climbaboard, Sugar Queen,* and *Come Get It*, were decorated with lush pictures of more or less undressed females, appropriate to their names. And one or two, I remember, had to be painted over or touched up, because the vividness of their newly acquired insignia interfered with the camouflage

—a distressing handicap to many an amateur artist in the combat areas!

Two days after Christmas, our awaited orders came through and we were on our way. Our destination: Apamama (spelled Abemama or Apemama on some maps), one of the Gilbert Islands seized from the Japs in the operation which resulted in the ousting of the enemy from Tarawa and Makin. The first half of the squadron left Kaneohe December 28; *Thunder Mug* and the rest departed a day later.

We flew by way of Palmyra and Canton Islands, an uneventful trip except that *Thunder Mug*, on the leg from Canton to Apamama, crossed the International Date Line exactly at the equator on December 31, 1943, an act which at 1:35 P.M. plunged us instantly into 1944 and deprived us of a New Year's Eve celebration. Once before I had been done out of a birthday; now it was New Year's Eve. The crew, solemnly sorry for their skipper, choraled *Auld Lang Syne* with appropriate melancholy. Later one of the boys made photographic copies of a hand-lettered card celebrating the occasion and passed them around. Mine reads as follows:

ANCIENT ORDER OF THE DEEP

NEW YEAR'S GOLDEN DRAGON SHELLBACK

Know Ye, By These Present That

N. M. MILLER, COMDR. USN

Having entered my August Domain and having crossed the International Date Line, the Equator, and entered into a New Year at one and the same time, at 0135 Greenwich Civil Time, 1 January 1944, while en route to Southern Waters in PB4Y-1 Plane No. 32108, should be recognized as a Flying Shellback, Devotee of the Golden Dragon, and a True Son of the Southern Cross.

BY ORDER OF NEPTUNUS-REX.

A few hours after crossing the Line, Bombing Squadron 109 was officially based at Apamama, ready for active operations against the enemy Japanese. But a word, first, about Apamama itself.

The atoll is a beautiful place, a few miles north of the equator and seventy-five miles south of famous Tarawa. About fifteen miles in diameter and shaped like a reversed "C," it consists of several small islands, all of them heavily capped with coconut palms. The gaps between these islands are fordable at low tide in a jeep, which fact was used to good advantage by the Marines when they took the place. The Japs had expected an assault on the largest island, and had constructed a defensive network of dugouts and trenches from which to repel the invasion. The Marines, landing on other islands, avoided the dugouts and trenches by attacking overland. Unlike the bloody affair at Tarawa, it was not a tremendous battle. There were twenty-six Japs on Apamama when we took it, and we killed one or two of them and lost one or two men in the process. All the rest of the Japs were shot through the head at close range, apparently by their officer, who then committed suicide.

The Japs as a matter of fact had not done much with Apamama. Perhaps they had not realized its significance. They had no air strip. Planes could not be accommodated until our Seabees moved in, cut down palm trees, brought in their draglines to scratch up the needed coral, and built one. The one they built was large enough—eventually—to accommodate two B-25 Army squadrons, a P-40 fighter squadron, a B-24 Army squadron, two Navy PB4Y bombing squadrons (officially "search and patrol"), and a Navy photographic squadron. From this Seabee air strip we were able to soften up the Marshalls for invasion while protecting newly won Tarawa and Makin with patrols.

In addition to the squadrons mentioned above and the 95th Seabees, Apamama supported a Carrier Aircraft Service Unit (CASU) and an ACORN, which is a mobile air station formed in the States and moved intact to the combat areas. For protection the island had a Marine defense battalion with antiaircraft and coastal defense guns. Yet the Japs had placed but twenty-six men on the island during their period of occupancy, and apparently these twenty-six Japs had done nothing but sit around and drink *sake*.

The Japs, while there, had tried very hard to conceal the fact that they were on the island at all. On one occasion, so I was told by a French priest who had been on Apamama for four years, an officer angrily shot a Japanese soldier who recklessly fired upon one of our reconnaissance planes. The same priest—a large, friendly fellow wearing a goatee and very black mustache—also told me that the Japanese had treated the island natives very badly. "Sometimes," he said, shaking his shaggy head, "when we were bombed by carrier-based airplanes, the Japs would storm through the village, searching house after house for the radios they were certain had guided the planes to the target. The natives had no radios, of course. But the Japs were terribly angry and expended their rage in brutal treatment of the people."

These Gilbertese inhabitants of Apamama were a friendly lot, childlike in their simplicity and eager to help. The men were husky and well built, the women and girls pretty and high-spirited. We found them quick to laugh, and soon learned that they liked us as much as we liked them, despite the fact that several had been killed by American bombs prior to the invasion.

The Gilberts, taken as a whole, consist of sixteen atolls which from the air look remarkably like a vast fleet of small ships spread across the "murmurous and sapphire solitudes of the Central Pacific." From north to south they are named as follows: Little Makin, Makin, Marakei, Abaiang, Tarawa, Maiana, Apamama, Kuria, Aranuka, Nonouti, Tabiteaea, Beru, Numunau, Onotoa, Tamana, and Arorae—and you can spell them almost any way you please, just as you can spell Soerabaja half a dozen different ways and still be right. Yet with all their wondrous names, the Gilbert Islands have a total land area of only 166 square miles, and none of them is more than half a mile wide or rises more than fifteen feet above sea level. Tarawa was devastated by bombing and gunfire before American troops went ashore to seize it. The others were left practically unscarred and remain as beautiful as Robert Louis Stevenson found them when he discovered there the answer to his search for rest and health. Stevenson found "a superb ocean climate, days of blinding sun and bracing wind, nights of heavenly brightness." We found the same, thanks in large measure to the Island Commander, Captain W. P. Cogswell, USN, who had the place really "in the groove."

Bombing Squadron 109's job at Apamama was primarily a search and patrol assignment. That is, each of our Liberators was to be a far-ranging pair of eyes with a radio set, alert to warn of any Jap counter-move in that part of the Central Pacific. VB-108, a similar squadron led by Commander E. C. Renfro, USN, was already there and had rolled up an impressive record. With their help, we pitched our tents under the coconut palms alongside theirs, set up our mess hall and offices in Quonset huts near by, and went to work immediately. On January 1, 1944, Lieutenant Commander John F. Bundy, while on routine patrol, came across a Japanese cargo vessel near Mille Island and sneaked up on it through a rain squall. He dived to 230 feet and dropped his bombs. When he left, after three bombing and strafing runs, the hapless Jap was on her way to the bottom and Bombing Squadron 109 had chalked up its first score.

Note, please, the altitude from which Bundy made his attack— 230 feet. This is "masthead" or "treetop" height bombing, and was to become VB-109's special pigeon in the months that followed. The men of 109 make no claim to having invented such tactics. Low-level bombing had been employed to good advantage in the Southwest Pacific, notably in the Coral Sea, and three experimental low-level missions by Liberator pilots in the Central Pacific were already on the books when we arrived at Apamama. Lieutenant Stickell, an ex-RAF pilot of VB-108, had already done much in pioneering that squadron in the techniques of masthead-height bombing.

In 109, however, we made a special and exhaustive study of the potentialities of this type of attack, and Lieutenant Commander Bundy, when he hit his Jap, was following to the letter a plan which we believed would prove to be enormously effective. Our line of reasoning went something like this:

1. Throughout the tour, our primary mission would be long-range search, patrol, and armed reconnaissance of the islands and

shipping lanes of the Central and North Pacific. This would provide many valuable opportunities for attacks upon enemy shipping and land installations, but in nearly all instances the plane involved in such attacks would be a long way from its base and would be alone. The best tactics, therefore, would be those of hit-and-run, and the element of surprise would be extremely precious.

2. In attacking enemy-held islands especially, the desired, all-important surprise could best be achieved by a minimum-altitude approach (20 to 50 feet above the water) to escape detection, using to the utmost the protective contours of the islands themselves. In other words, the attacking plane was to hug the sea in a swift approach, rise suddenly over intervening cliffs or obstacles, and come in upon the enemy at high speed before he could be aware of his peril.

3. Bombing would be done on a precision, pinpoint basis from an established low altitude, at predetermined speed, with all ten .50-caliber machine guns blazing away at the same time to prevent the enemy from manning his defenses. Then, its mission accomplished, the attacking plane would "get the hell out of there" as quickly as possible, before the enemy could organize an effective pursuit.

These, roughly, were the tactics we had worked out and diligently practised from the day of our commissioning. How well the method worked will be revealed in the following pages.

13   *PRELUDE TO INVASION*

OUR elation over the squadron's first score was dampened the following day when the Japs bombed Apamama. They caught us at a bad time, with parking facilities still inadequate and the squadron's planes jammed on the taxi strip. Had they been more accurate with their bombs, or in less haste to get home again, the damage might have been paralyzing. As it was, an incendiary bomb set fire to Lieutenant George Mellard's plane, bomb fragments damaged three others, and three men were hurt.

More serious damage was averted when four men—Ensign Amos Shreiner, Ensign Bob Malmfeldt, Harry Donovan and Dan Dujak— risked their lives to move other planes from the vicinity of the one set afire. You think twice, or ought to, before fooling around a blazing airplane loaded with high octane gasoline, 500-pound bombs, and .50-caliber ammunition, but these men rushed in. The burning plane, totally destroyed, was our only loss, though explod- ing ammunition shattered the windshield of one plane as Shreiner taxied it away. Luckily he was not hurt.

"Now, dammit," said Bob Malmfeldt disgustedly, "I'll be having nightmares again. Did I ever tell you what happened to me on Tulagi? No? Well, some Negro Seabees were working there one night with lights strung on the trees, when a Jap bomber on the way to Guadalcanal picked the place for a target. No one heard him coming; he had cut his engines and made a sneak approach. He dropped at least five bombs, and not one of them landed more than a hundred feet from the tent where I was sleeping. A lot of men were killed. A number of others were injured. As for me, the explosions blew me so violently through the mosquito netting that I had nightmares for weeks afterward. About a week later a fellow slammed a head door a few feet from where I was sleeping, and when I woke up I was running at top speed down the corridor, yelling, 'Hit the foxholes, you guys! Hit the foxholes!' Those," Bob concluded with a shrug, "were the days when the Japs really laid eggs on us."

We didn't argue, but we thought the Japs were doing well enough at Apamama, too. Fortunately they were jittery, and after facing our AA fire again the following night without doing any real damage, they left us alone.

Thunder Mug's first patrol assignment was on January 3, when three planes of 109 and four of 108 carried out a joint mission over Maloelap Atoll. Maloelap lies in the Marshalls, and with Kwajalein, Wotje and Jaluit was considered one of the fangs of that Japanese dragon's-mouth.

No one seemed to know very much about the Marshalls. At the

conclusion of World War I they were wrested from Germany and handed over to Japan under League of Nations mandate, whereupon the Japs drew a curtain of secrecy about them, ruthlessly shifted the native populations around, and expelled all Europeans. We did know, however, that several of the thirty-two Marshall Islands were the sites of major airfields and submarine bases, and our job was to keep an eye on these islands, heckling them at every opportunity, while preparations were being made for invading them. The trip to Maloelap—*Thunder Mug's* first—was part of this program.

I remember the occasion not for the damage we did but for an incident that occurred as with *Thunder Mug's* crew I traveled in a jeep down the road to the air strip. The road was narrow. A procession of cars, followed by a truck, forced us to move aside. In the truck lay a coffin containing the body of a Marine AA gunner killed the night before. We sat quietly, hats in hand, as the funeral cortège passed, and then without comment continued on to the plane. The lump in our throats was not easily swallowed.

A few nights later that Marine was partially avenged, when Lieutenant Commander Janeshek, Lieutenant Jobe and I, with four planes from 108, hit the island of Wotje on another special mission. It was a dusk raid. As we approached the island in line abreast, traveling the last hundred miles just above the sea, the setting sun was golden bright and the island's palm trees waved like colored feathers in the breeze. Bright, too, were the Jap installations, especially the hangar at the far side of the airfield.

The hangar doors stood open; we had caught the Japs by surprise. As *Thunder Mug* swept in with her machine guns spitting, frantic little figures scurried about inside the building to escape the stream of bullets. Many crumpled as they ran. Then we were past, and the structure was blazing behind us, and we felt a little better about the funeral procession at Apamama. In the morning there would be a funeral on Wotje, too. A big one.

The Japs' major base at Kwajalein was hit a few days later, this time boldly in broad daylight, by eight Liberators from 108 and 109, and two from a Liberator photo squadron, VD-3. We had studied all available photographs the night before, at a prolonged briefing, and the plan of attack was carefully drawn up. Nothing was left to guesswork.

The more important part of Kwajalein Atoll is shaped roughly like a V, open at the top, with Kwajalein Island, our target, at the extreme bottom. Six of our Liberators were to cross the island from the south, in line, selecting the largest targets of opportunity among the shipping in the harbor beyond. Two planes were to drop back and cover the flight on the run in. All planes would then turn sharply to the left and retire across the left shaft of the V, near South Pass, the two guardian planes unloading their bombs on Enubuj Island to hold the Japs down. The photographic planes, of course, would take pictures.

Our flight from Apamama was leisurely and in good order. Now and then we encountered rain squalls, but most of the time the sun was dazzlingly bright overhead and only a few small clouds colored the sky. Entering the Marshalls between Jaluit and Mille (spelled Mili on some maps), we went down to fifty feet when still fifty miles from the target, and increased speed as Kwajalein came in sight. The surprise was perfect. Before the Japs knew what was happening, six of the big four-engined bombers were thundering in line above them, with all guns blazing. *Thunder Mug* was in the middle of the line, and so low that I could not see over the island's palm trees to spot the ships in the harbor beyond.

There must be something terrifying in the sight of a fleet of huge high-altitude bombers hurtling toward one at treetop level. They are so unbelievably big, and make such a racket! At high altitude they can be regarded with a certain amount of detachment—if a bomb from that height hits you, then perhaps an element of fate is present. But at fifty feet, just over your head, the big planes must seem a good deal more personal, as though determined to single out you and you alone as they come roaring in. This is pure speculation, of course, but on Kwajalein that day the Japs scattered as though pursued by all the private devils of their Shinto creed. They ran in every direction to escape destruction.

Their trouble lay in finding somewhere to run, for with six

planes abreast, spaced about one hundred yards apart, the rain of
.50-caliber bullets that swept the island was really a deluge. I saw
little of the action myself—at 200 miles an hour a pilot is far too
busy watching for new targets—but the crew later recalled many
snap impressions. "That guy's legs crossed and he skidded ten feet
on his face." "That Jap just leaped into the air and, so help me,
didn't come down. He exploded."

We had passed over the narrow island in a matter of seconds,
confining our attack to strafing, and were suddenly out over the
lagoon, two hundred feet above a covey of cargo ships anchored
there. I made my run on a 5,000-ton cargo vessel, pickling * off five
bombs as the ship loomed in the sights. Two were short, two were
direct hits, the last was over. There was a huge explosion, clearly
shown in the pictures taken by the accompanying photo planes,
but I failed to see it, for by then the enemy AA had opened up
on us and I was "jinking"—that is, violently swerving the plane left
and right—and seeking other targets. But as we flew on over the
lagoon, the smoke cleared and the tail gunner gleefully reported
that our cargo ship was blazing beyond control. Later, from a distance, I saw the flames myself.

That was not the only damage the Japs suffered. Before our six
Liberators had swung left to retire across Enubuj, Hicks and Grayson had scored hits on enemy shipping, and Lieutenant Oden Sheppard, whose assignment called for an attack on the air strip, won
himself a Distinguished Flying Cross by bombing and strafing eight
Zekes (Zeros) on the runway and taxiway, two of which were already turning up for a take-off. The score, for VB-109 alone: two
planes certainly destroyed, three cargo vessels and six planes probably destroyed. And we had scared the pants off a few thousand
Japs who probably never had dreamed that they might be attacked

---

* On all but one or two of *Thunder Mug's* missions the bombs were released by the pilot himself, by means of a gadget known as a "pickle"—so
called because of its shape. Having predetermined the interval at which the
bombs shall be dropped, and set the intervalometer to attend to this, the pilot
holds the pickle in his hand and presses a button on the end of it to release
the bombs from the rack. Hence the term "pickling."

in broad daylight. From now on they would be hampered in their activities by the need for being everlastingly on the alert.

Once again, however, the Japs found a way to temper our triumph—this time seriously. Two days after the Kwajalein strike, Lieutenant Sammy Coleman and his crew took off on a routine search and were not heard from again. Over a four-day period the entire patrol sector was searched again and again by planes of our own and other squadrons, but no clue was uncovered. The missing Liberator had simply vanished.

It was a saddening blow, for by this time we were a close-knit group. Co-pilot of the missing plane was Amos Shreiner, the man who had flown with the RAF on twenty-one missions over Germany before joining us. Among the crew were Harry Donovan and Dan Dujak, who had risked their lives to save our planes the night of the Jap attack on Apamama. Reluctantly, after four days of intensive search, the plane and crew were listed as "missing in action" and the search was abandoned. The Pacific holds many secrets.

Thunder Mug, of course, took part in this search for Lieutenant Coleman, and while ranging through the Marshalls north of Kwajalein one day I thought it might be wise to look in on Likiep Atoll, which lies about midway between Kwajalein and Wotje and a bit to the north. We knew little about Likiep and had never been there. A look at the place, I felt, would provide valuable information for Lieutenant Ed Lloyd, our Air Combat Intelligence Officer (ACIO), as well as the assurance that we had overlooked nothing in the search for the missing Liberator.

It was not much of a place, and apparently most of the tiny islands which made up the atoll were uninhabited. On the west side of the most southeasterly island, however, I discovered a large patrol ship lying at anchor, just off what appeared to be a small, inhabited village. The ship was Jap, with the white numerals "406" painted on each side near the bow in figures four feet high, and a three-wire radio antenna strung between her masts, eighty feet above the deck. That she was a steam vessel was indicated by the stack amidships.

We flew over her at masthead height, noting the presence of guns, apparently 3-inch, on platforms at the bow and stern, trained upward toward us and covered with a palm-leaf camouflage. No one appeared on deck to man them as we came about and dropped two bombs.

It was strictly a "milk run." But as so often happens when the skipper is putting on a show for the boys, the results were not so good. Both bombs were over. Chagrined, I pulled the plane around for a second run and laid two more bombs, all I had, on the same target. One was a direct hit. Where the other one landed we never knew, for the Jap blew up with a mighty explosion, and by the time I was able to circle again for inspection, nothing was left except floating debris. We departed without strafing the village—the inhabitants may have been native islanders—and without, unfortunately, finding any trace of the missing Liberator.

That same day, Lieutenant Commander Janeshek, also searching in the Marshalls for the missing plane, got himself a Jap ship near Jaluit, to add to our total. Despite the fact that the area had been well worked over by VB-108 before our arrival, and the wary Japs were keeping their more vulnerable shipping carefully hidden, our score was going up. But we were paying a high price for it. Lieutenant Coleman was gone, and on January 27 Lieutenant Jadin, in attacking two ships south of Eniwetok, suffered severe damage to his plane and brought home a severely wounded tail-turret gunner. The targets we were finding were potent. Only the fact that we were piecing together an important amount of valuable data on enemy installations in the Marshalls—information which would be used by our invasion forces—made it worth while.

The invasion itself took place shortly afterward. On the last day of January, Lieutenant Seabrook, flying patrol over the western Marshalls, witnessed the shelling of Kwajalein by American naval forces. The attacks had been initiated by carrier-based planes the day before, and were continued by carriers and warships through February 1. Troops landed on Majuro on February 1, on Roi,

Namur and Kwajalein on February 2. Stiffest resistance was encountered on Kwajalein, which fell to our troops on the eighth day of the month. Other important islands in the Marshalls—Taroa, Wotje, Jaluit, and Mille—remained in enemy hands, but at least we had obtained a foothold for expansion.

[*For these early raids on the Marshalls, Commander Miller received his first and second Air Medals, with citations reading substantially as follows:*

Air Medal: "*Participating in vital bombing and strafing raids on hostile shipping and shore installations in these strategic islands prior to our invasion operations, Commander Miller skillfully maneuvered his plane at extremely low altitudes over his targets in the face of heavy antiaircraft fire, carrying out effective, accurate strafing runs and inflicting severe damage on the enemy. On one occasion, while engaged in a routine search flight, he discovered a Japanese ship anchored in a lagoon and, pressing home his attacks, completely destroyed the hostile vessel.*"

Second Air Medal: "*As a Patrol Plane Commander in action against enemy Japanese forces in the Marshall Islands on January 11, 1944, Commander Miller courageously led a division in a daring, aggressive daylight strike on hostile shipping in Kwajalein Harbor, pressing home his powerful attacks at perilously low level in the face of persistent antiaircraft fire, scoring two direct hits on a hostile cargo vessel which was left burning and in a sinking condition. Commander Miller's expert airmanship and indomitable fighting spirit, maintained at great personal risk, were in keeping with the highest traditions of the U. S. Naval Service.*"—H.B.C.]

14   *THIS IS WHAT IT'S LIKE*

LIFE on a small Pacific island can be dull, even in a well-kept community. On Apamama, where our job primarily was to learn the ins and outs of a new business, we tried to be as comfortable as possible despite the handicaps, and we tried to keep the job itself interesting. Long flights day after day can be boring. Some of ours were very long indeed.

I soon learned that the squadron bulletin board, for instance, was not getting the attention it deserved. Like most such boards, it had accumulated a mess of routine papers which the men simply ignored. So the name of it was changed to the "Red Hot Dope Board," and a rule was inaugurated that nothing, no matter what, was to remain on it for more than three days. To stir up interest, I posted jokes, cartoons, anything that came along. (This continued throughout our tour of duty, and before long we were getting all sorts of lively items from the folks back home. Thelma, for instance, used to include jokes and newspaper clippings in almost every letter.)

Of course, it worked. A new joke, a new cartoon, becomes an item of major importance in a small community of men who have heard one another's stories over and over and are desperate for something fresh. The board was no longer ignored, and I never again allowed it to become cluttered with needless orders. Only the essential ones were posted. Other information was passed out verbally, when we got together to discuss the "dope."

These discussions, held in the skipper's tent, the mess hall, or the ACI office, often lasted for hours. Pilots, co-pilots and navigators, drifting in and out, would gather around to examine maps of the islands which were likely to be our targets. Ed Lloyd, our ACIO— before the war he had been a Pasadena lawyer—would give us all the information he had been able to pick up concerning these islands, telling us what the enemy had on them, which were strong, which were weak. Then, quite informally, the boys would go on to discuss ways and means of getting in to make trouble for the enemy.

My own particular pet was the Jap stronghold at Truk, and I carried with me a little yellow-covered booklet of Truk target maps. Long before I ever ventured into the place, I did my best to know the names of the Truk islands by heart, what they would look like from the air, where the enemy had set up his major defenses, and where his anchorages were. On Apamama we were a very long way from this Jap bastion, of course, but eventually we would be moving our base into the Marshalls, and then Truk would be within reach. Some day I was going to find out what Truk really looked like—not on maps or in photographs, but the real thing.

Through January and most of February, however, our flights continued to be more or less routine. Not that they lacked their moments. No flight is ever completely boring. We took off at all hours, usually hitting our targets at dawn or dusk, or at some selected hour when the moon would be of use to us in silhouetting certain items of interest. Most of these targets were at extreme cruising range, which increased the element of danger. For a dusk strike, we left Apamama about noon; for a moonlight strike, sometime in the afternoon; for a dawn attack, about midnight.

A strike, let's say, has been scheduled for dawn tomorrow at Island "X"—a tiny place with an unpronounceable name well up in the Marshalls, where reconnaissance planes have reported some important enemy shipping. *Thunder Mug* is handed the assignment. Old Shafe *—Lieutenant (jg) Robert K. Schaffer, my navigator—fusses with his charts and tables for a while, and together we figure out the take-off time. Shafe, a tall, extremely good-looking boy with jet black hair—until it turned gray—used to worry a lot when we were scheming up these missions, but he would go anywhere and the crew worshipped him. In fact, I twice had to remind them gently that we were still in the Navy where it was considered not quite proper for enlisted men to address an officer as "Hey, Shafe!" "*Mister* Schaffer," I used to say, wagging my head. And "Mr. Schaffer" it would be, but I suspect that it became "Hey, Shafe" again more often than not when my ears were turned. Shafe or Mr. Schaffer, he was one of the finest navigators in the Navy, and if Old Shafe said you would be off the reef at Island "X" at

---

* *The word "old," as used by Commander Miller in speaking of "Old Shafe," "Old Thunder Mug," "Old Ted," etc., has nothing to do with age. It is a term of affection, as characteristic of the commander as his favorite cuss word, "Hellacious!"—H.B.C.*

0502, you'd be there at 0502. Nor did I ever have to vary our course on the way back, to pick up the home island. Nine times out of ten, even after fourteen or more hours in the air, Shafe would bring us home with such accuracy that with only a wiggle of the wings, and sometimes not that, we could sit right down on the runway. Whereupon I usually turned to him and said, "Shafe, you're the lousiest navigator in the business." And Shafe would say, "Skipper, I can't help it if you don't fly the way I tell you to." And I'd say, "Shafe, you haven't stopped shaking yet." And Shafe would say, "Skipper, I'll never stop shaking. I'm scared stiff. But I can go anywhere you can, by God, and I'll prove it."

Old Shafe, then, has figured out our take-off time, and I drop in at the operations office to tell Lieutenant "Buzzy" Robinson, our Operations Officer, where we are going, how much gas we'll need, and what sort of bomb load we plan to tote. Then to Ed Lloyd for the latest intelligence on Island "X." Then, armed with maps and charts and all the available dope, we pick up the crew and go down to the air strip to pre-flight the plane—that is, turn up all the engines to be sure they are sweet, see that the radioman has checked his radio equipment, and give the entire plane a thorough once-over. We are going to be in this plane for fourteen hours or more, much of that time over an empty ocean or enemy-occupied islands. Only a very careless crew would embark on such an undertaking without first making certain that to the best of their knowledge the plane was mechanically perfect.

With all the crew present, we get together under the wing, now, for a briefing. "Boys," I say, unrolling the maps, "we're going in to Island 'X' tonight. Here's how we'll do it." With the maps unrolled on the ground, I give them complete information on how we are to approach the island, the targets we'll aim for, the enemy defenses to look out for, and the route by which we will retire. I tell them, too, the details of the alternate plan which is always held in readiness in case something goes wrong. Everything, after all, depends on the crew. They cannot be at their best unless they know what to expect, and if they aren't good, the pilot is simply flying an airplane.

(This was a lesson I learned on one of our early missions, when I purposely neglected to tell the crew where we were going. Upon landing, they walked away in silence, without any of the usual chatter or clowning. I asked them what was wrong. "Skipper," they said, "we'll go anywhere you will—to hell and back, if you say so— but we want to know about it beforehand. We don't want to be treated like scared kids." Well, some of them were kids, young enough to be my own, almost; but they were the grandest crew a pilot was ever blessed with, and they certainly were not scared. I never again neglected to give them the dope, all the dope, before we took off on a mission.)

So, then, the plane is ready, the crew has been briefed, and we have nothing much to do until take-off time. If it is to be a difficult mission—if, for instance, Island "X" is known to be strongly defended—there is likely to be a little tension about now, though you must look closely to spot it. One man in whom you won't find it is Paul Ramsey, the plane captain—senior enlisted man of the crew— who has no visible nerves whatever and is concerned only with the plane, which is his responsibility. (Secretly, perhaps, he was concerned also with a Jacksonville WAVE, whom he married soon after our tour of duty ended. A competent, soft-spoken boy from York, Pennsylvania, Ramsey knew all there was to know about his plane and is now an ensign, after working his way up in the regular Navy.)

Some of the boys, however, are clowning about, hiding their tension under good-natured kidding; others are soberly discussing the mission and trying to pick holes in the plan of attack which has been outlined to them. They are smart; you don't fool them. If the plan is not perfect, they will find the flaws.

Meanwhile, I have gone back up to the ACI office to be briefed myself by the ACIO, taking with me the co-pilot and navigator. (If more than one plane is to be in on the mission, other pilots, co-pilots and navigators will have gathered there as well.) From Ed Lloyd we get the latest information on enemy shipping, the location of friendly submarines if any are likely to be encountered along our track, and any other intelligence that may be of use to us. We take notes. We discuss the attack from all angles, make suggestions, finally reach the point where the subject is quite talked out.

It is now late afternoon and we have been working on this mission all day. I hand the co-pilot or navigator the briefcase containing my notebook and the charts. He takes it to the plane. The navigator takes along, too, his octant and gear and his own charts, well worked over by this time. We are then at liberty to go to a movie or turn in for a few hours' sleep until take-off time.

To be sure that I won't oversleep, I set my alarm clock and direct the duty officer when to wake me. But sleep is elusive at such a time. Instead, I lie there thinking. Again and again I go over the plan of the mission, seeking flaws in it, wondering if this time I have overlooked some small but vital item which may mean the difference between success and failure, perhaps even between life and death. We know so little about this Island "X" really. What if the Japs have moved that potent AA battery to a position nearer the lip of the channel, where it will blast us as we come up off the water to clear the cliff? What if they have put a lookout on that little outlying island over which we are planning to fly to reach our target?

It is 12:30 when the alarm goes off. The tent is dark. Sitting up, I reach sleepily for clothes and shoes and become aware of the sound of the surf on the island's reef—funny, but you never notice that sound in the daytime; you hear it only at night, when all else is still. At the mess hall, which is a Quonset hut, the co-pilot and navigator are waiting. They, too, are rubbing sleep from their eyes. The crew, up ahead of us, have gone down to the plane.

It's a fair breakfast: fried eggs, the inevitable spam prepared in some trick manner, and coffee or a glass of water. The co-pilot has picked up the flight rations—sufficient food to keep the entire crew going until we return to base—and carries these in a cardboard box under his arm as we climb sleepily into the jeep that is waiting to take us to the air strip. Under my arm I carry a wadded-up pair of coveralls, the pockets stuffed with a strange assortment of objects: a steel signaling mirror, morphine syrettes, a small pocket compass, dog tags, a long knife with an 8-inch blade, a canteen of water, a pistol and ammunition, and my rabbit's foot. At the plane I strip off my clothes and don the coveralls; it is easier to leave the clothes behind than to go through all the pockets in search of bits of paper that may contain confidential information. Because, of course, we may not come back. A Liberator is a good airplane, but it is only an airplane, not a magic carpet. And the Jap AA gunners and fighter pilots are sometimes very competent people.

The members of the crew are waiting on the air strip, all of them sleepy but on edge now, eager to be in action. I give them the last minute dope if there is any. Someone—usually it is Bobby Gariel or Thomas Delahoussaye—has a joke to tell or a crack to make. "Now listen, Captain. If those flying fish get into my turret again and spoil my aim, will you *please* pull up to the curb till I bail them out?"

But Old Shafe says it is time to go. "Okay, let's crank her up." Ramsey, the plane captain, winds up the engines, listening intently to the song they sing. Bobby Gariel, the San Antonio tail gunner, climbs up through a waist hatch and goes aft to his tiny cubicle in the tail, stepping over the little metal "head" can en route. Bobby is dark and wiry, with broad shoulders but disgracefully narrow hips. He fits nicely into the cramped quarters of his tail turret, and while he sits, waiting for action, he thinks up most of the wisecracks for which the crew of *Thunder Mug* is famous. Such as the day he solemnly explained the enemy's failure to halt one of our more audacious attacks on a Jap stronghold. "Sure they saw us coming. They had lookouts all over the island. Trouble was, those Japs rushed to their phones and babbled the alarm in *Japanese*—and who the hell understands Japanese, anyway?"

Eddie Dorris, the eighteen-year-old from Kansas, tall, thin, always a gentleman, shrugs himself up through the hatch to the flight deck and checks his radio equipment. Or, if it happens to be his turn to take the top turret, where he alternates with Wayne Young, he just continues on up, threading his thin body through the turret hatch into the glass bubble above. Despite his youth, he is a remarkably mature kid, keenly observant. He is a particular favorite of the

ACIO, who can rely on him after a mission to report quite detachedly on what happened, even to the extent of criticising the skipper.

Wayne Young is that way, too. Twenty-five or twenty-six years old, from Seattle, he is the granddaddy of the crew, an accomplished radioman. His clowning is of the quiet kind; his jokes are more subtle. He takes his place now at the radio, in a two-by-four niche just behind the cockpit, where, with the headphones on, he makes a few last-minute checks and adjustments.

Lawrence Johnson, the belly-turret gunner, is a big lad from Grampion, Pennsylvania, where he used to be a lumberman. Lumbering was good for Johnny. He is ruggedly built, strong as a horse. But working in the woods has made him unduly quiet, too, and though he is good-looking, with an affectionate small-boy sweetness when he smiles, his face in repose is often heavy and serious. You have an odd desire to mother him—but how in heaven's name can you mother a husky brute with a physique like that! Johnny has no nerves; he is quiet, thoughtful, and no one ever knows whether he is scared or not. Yet he must be scared at times, for his job in the belly turret is nerve-racking. Hour after hour he must squat there in his glass bubble beneath the plane, watching the ocean skim past below him as we fly at high speed just over the water. If the plane should falter at that speed, or if the pilot should make a mistake, Johnny's little bubble-world would be wiped out in an instant, swept away like a drop of water from the belly of a skimming duck. Many a time he picks up spray from the waves as we go rumbling across the sea.

He stands by, now, waiting for the plane to be air-borne before squeezing his big bulk into the blister. Then the blister will be lowered hydraulically and he will be in a world of his own until it is hauled up again—a world connected with the rest of the plane only by interphone. The boys don't envy Johnny.

They don't envy Jack Simmen, the bow-turret man, either. Simmen is a big, quiet boy, an excellent gunner, but when we make our low-level attacks his turret is a natural target for the enemy's AA fire. He has calm nerves, and he needs them.

Delahoussaye and Jaskiewicz, the waist gunners, have more space than anyone else in which to stretch their arms and legs. Both are good boys. Both will probably have burned out their gun barrels by the time we get home again, unless they run out of ammunition before that happens. Delahoussaye, from Lafayette, Louisiana, has some French in him and looks it. He is fountain-pen thin and filled with good humor. Jaskiewicz, who hails from Chicago, is precisely what you'd expect with a name like that: big, brawny, tough, incredibly cool. "Whisky Jazz," the boys call him. His heart is as big as the Pacific sky through which presently we'll be flying to our target.

They are all aboard by now: tail-turret gunner, belly-turret gunner, top-turret gunner, bow-turret gunner, radioman, the two waist gunners, plane captain, navigator, co-pilot and pilot. The engines are turning up. In the cramped, crowded little cockpit the co-pilot (I had no regular co-pilot then, and borrowed men from other crews) occupies the hard, postage-stamp seat on the right, and the pilot has squeezed himself down on the one at the left. Between us stands the plane captain with the check-out list, reading off the items to be checked before we "haul the hook." Nothing is left to memory in handling these big planes. Before every take-off and landing, the list is gone through to be sure that all the complicated flight and control instruments have received the proper attention from pilot and co-pilot. I counted those instruments one day. For your information there are 204 levers, switches, gadgets and push-buttons to handle and 60 indicators and dials to read. And I probably overlooked some of them.

We complete the check-off list, turn up the engines and set the turbo superchargers. I turn to Ramsey, the plane captain, with the routine question, "Everything okay?" He nods, and we taxi out, following a jeep down the runway to the take-off position. There is a big sign on the back of this jeep at Apamama. "FOLLOW ME," it says. And so our huge, lumbering, four-engined bomber meekly trails a toy automobile, while the plane captain holds his head out of the top hatch to be sure the path is clear.

Old Shafe, worrying as usual, solemnly looks up from his navigator's charts. "Skipper, I hope you're right about those AA guns on the point." I hope so too, and grin at him—the boys have told me that no matter how concerned they are about our chances, everything is all right when the skipper smiles. I sometimes have to dig deep for that smile.

With the engines bellowing, we watch the tower for the green light, and at last we get it. *Thunder Mug* rolls down the runway, straining to be air-borne with a load of gas and bombs that would cause a lifting of eyebrows at the factory where she was built. The island drops away in the darkness beneath us. We come about, straighten away on course, and begin the long, lonely flight through the night to our destination. And because it will be many hours before we are able to get out and walk again, each member of the crew begins then and there to apportion out his stored-up energy in calculated dribbles, to make it last. Few words or motions are wasted.

It is a strange and sometimes awesome experience, this flying over an island-dotted ocean at night. There may be a moon to silver every wave-top and trace patterns of extraordinary beauty as far as one can see. There may be no moon at all, and then the sea and the islands are blacked out except for oddly distorted shadow-shapes that well up in the darkness and are gone again—islands like black bubbles turning up from the depths. But sometimes there are no islands. Sometimes for hundreds of miles, even thousands of miles, there is no land at all, anywhere, and then the Pacific is large and lonely.

We are tense, and the tension grows as the distance to the target decreases. The hours drag interminably. I see the co-pilot rubbing his thighs to increase the circulation, and I rub my own, always watching the sea and waiting for the emergence of the next ghostly cluster of islands—if any—along our track.

Beside me on the deck of the cockpit is the chart I always take along, folded so that the portion covering tonight's flight is uppermost. Old Shafe, facing forward in his navigator's compartment, holds a flashlight ready, and when I say "Light, Shafe," he directs a beam downward and I check on a passing island. "We're okay. That was Namorik. Wasn't that Namorik, Shafe?" But it has to be done quickly, for as we near our destination we are very low over the water, turning up tremendous speed, and if my attention is diverted from flying for more than a second, we may be in trouble.

Yet these quick glances at the chart, especially at the target chart, are highly important when approaching an enemy atoll, for the islands loom up with bewildering rapidity as the target is approached. Then in a matter of seconds the pilot, to orient himself, must make a lightning transition from lines on a chart, seen vertically, to the islands themselves, seen horizontally. And he hopes to heaven, as he does this, that Island "X" will look enough like the map, despite the darkness, for him to spot his objectives.

Now, with but a few more miles to go and the sky beginning to lighten in the east, we are very tense. Just above the surface of the ocean, old *Thunder Mug* roars along at 200 miles an hour or better, as I reach for the interphone.

"Okay, boys, we're going in just as we doped it." No need to tell the crew to be on their toes. They, too, are tense.

Suddenly the islands are upon us, and if we have come in from the west, to silhouette them against the dawn-light, they loom huge and blurred and black above the plane. We are skimming the water —a few feet lower and Johnson, in his belly turret, would find himself in a diving bell. We rise sharply as the islands thicken up about us. And "thicken" is the word. They seem to rush in from all directions, for these mid-Pacific atolls are usually composed of many little land-masses grouped in helter-skelter fashion around a central lagoon.

We know our target. Our course has been charted so that we will hit the maximum number of enemy installations: airfields, ammunition dumps, grounded airplanes, radio towers, or whatever the Japs have there that we know about. And always, of course, the shipping. For the Japs are short of ships, and these cargo vessels and their escorts are the lifeline by which their outlying island bases draw sustenance from the homeland.

So far we are not seen; the priceless element of surprise is still ours. The gunners hold their fire until the co-pilot, watching out the right-hand window of the greenhouse, begins to warn them of targets coming up. That is part of his job, to keep the crew informed over the interphone. "Ship on our starboard bow!" he warns. "Watch it! Planes parked on the apron to port!" Thus warned, the gunners can be ready. Bobby Gariel in the tail, for instance, can train his turret around so that when we roar over the ship in the lagoon, he need only catch it in his sights and blaze away at it.

But the co-pilot must be very good at this sort of thing. He must be hair-trigger quick as the targets come rushing at him, and he must be calm and sure of himself, especially sure of his voice, lest he rattle the men at the guns. The best of all co-pilots was Lieutenant Kasperson, who used to keep up a quick but quiet flow of talk that went something like this: "Delahoussaye, your side, couple of Bettys on the ramp ahead . . . Watch your fire in the belly turret, don't hit the bombs . . . Whisky, get that so and so in the tower . . ." Old Kas worked the gunners surely and confidently, never confusing them. They were seldom so effective as when he stirred them up with his matter-of-fact little pep talks.

So we go in, and the tension which seemed unbearable up to the moment of attack now becomes sheer excitement. Bombing and strafing, we sweep across the target island, our plane lit up like a Christmas tree as the gunners "give 'em hell" at my signal. We are spouting fire from ten guns, an awesome spectacle, and can see our own tracers ricocheting from the ground—sometimes they ricochet so viciously that *Thunder Mug's* own belly is scarred as we travel through them. And the Japs, if they are alert, are filling the sky around us with their own tracers, easily distinguished from ours because they are a weird greenish blue. Some of them may hit us. Bigger shells, too, may hit us as they come hurtling up from the stronger gun positions, bursting into gaudy blue, red and green flares. But a Liberator can take a lot of punishment.

Often the damage we do is not fully determined, for at treetop height we must use delayed-action bombs to avoid blowing ourselves up with the Japs, and in the four or five seconds before a bomb explodes, the plane has traveled almost a quarter of a mile.

The pilot sees almost nothing; he is usually jinking to avoid enemy counter-fire. The waist gunners and the tail gunner may observe the results of our attack if they are not too busy blasting other targets, but their reports are usually brief. The irrepressible Gariel, in the tail, may shout over the interphone, "Beautiful, Cap'n!" or someone else may groan, "Dammit, we missed!" but most of the talking is done later. The job now is to do all the damage possible while our bombs and ammunition last—to smash the air strips the Japs have so laboriously built, sink their ships, burn their planes, knock out their radio and radar equipment; in short, to liquidate these island bases which block our sea-roads to the Japanese mainland. To hell with looking back at what you've done! Look ahead and do more!

And then, before the enemy's AA fire becomes too hellaciously hot, and before his fighter planes take off to knock you down, "let's get the hell out of here—fast!" That was the phrase we used. That was our sign-off. "Let's get the hell out of here!" Later it became the chorus of the squadron song.

So we get the hell out of there—all in one piece, if we've been lucky, but sometimes limping home on three engines, with old *Thunder Mug* offering a fair imitation of a sieve. Perhaps we are chased by enemy planes, and then, full throttle, we hunt for cover in rain squalls or clouds, playing hide-and-seek until the enemy abandons his pursuit.

At last the crewmen relax. The flight rations are handed out. The radioman sends in a coded flash report of what we discovered this time at Island "X"—a report which passes on through many hands and is studied eventually, if important, by officers of all the air and sea forces operating in the area. The co-pilot takes over. Old Shafe checks the homeward course, so that in the end, after many more miles over that very big ocean, we will hit Apamama on the nose. (And right here let me salute Schaffer and all his kind, for the whole essence of a mission such as the one just described depends on the navigator's ability to place the plane over the target on the right heading at the right time. To Captain Woodruff and Lieutenant Commander Brownell, who established and conducted the Naval Air Navigational School whence these navigators came, unlimited credit is due. And to the boys themselves, many of whom could have been pilots had they not volunteered to study navigation instead, my hat is off in tribute. I have always felt that Navy navigators should wear some special insignia to distinguish them. It's the navigators, God bless them, who get you there and bring you home.)

It is Shafe who brings us home this time, and as *Thunder Mug* rolls to a stop on the Apamama runway, tired men drop stiffly to the ground to stretch arms and legs which have been aching for hours. Twelve, fourteen, sixteen hours in the air is a long time. We took off at 1 A.M. It is now 3 P.M. All that time in the air for twenty minutes over the target! "If I had to sit that long on a day-coach," says Shafe, "I'd sleep for the next two days." But a Liberator, despite its size, is far more confining than any day-coach. Pilot and co-pilot are so hemmed in by instruments that often they stay glued to their chairs the whole way, rather than go through the calisthenics of extricating themselves. If they do "go for a stroll," they must clamber down from the cluttered cockpit to the catwalk, a strip of metal less than a foot wide, thread their way along this between the bomb bays, and wriggle around the belly turret-well to the waist before there is room for anything more than light breathing.

Strolling under such conditions is not only difficult, it is dangerous. I tried it once when we were limping home with our hydraulic system shot up, and stepped without thinking on a catwalk made slippery by a spray of oily red fluid from the shattered hydraulic pipes. If the bomb bay doors had been open, I would not be here. I tried it again on the way home from a mission during which we had jettisoned the belly turret to lighten our load—and a sudden lurch of the plane sent me stumbling toward the open well. Grabbing at everything within reach, I hung on for dear life, scared stiff, as the dark Pacific rushed past below.

But at any rate, we are home, eager for chow and sleep. We don't sleep, though. Not yet. Ed Lloyd, our ACIO, is hungry for information and pounds us with questions. His job is to find out what happened, all that happened, to check and double-check our stories so that there may be no mistakes in the combat report which he must write.

Then, but not until then, the mission is over.

"A lone Navy Liberator penetrated the defenses of Island 'X' yesterday, dropping bombs on enemy installations and damaging enemy shipping in the harbor."

15   *"ONE OF OUR PLANES IS MISSING"*

SOME of the correspondents have called it "Miller's Secret Bombsight." Actually it was not a bombsight at all, and the only secrets involved were the speed and altitude at which it was used. It developed out of a realization that the results we were getting were not the best that could be obtained. In other words, we were not hitting enough of what we aimed at.

In a purely experimental mood one day I took off from Apamama in *Thunder Mug*, with Lieutenant (jg) Johnny Herron as co-pilot, and went looking for an island where we could do some practise bombing. We found a suitable place not far distant, selected a spot on the coral beach for a target, and began making runs over it.

I had been pondering the problem for days, and had come up with a hunch that by using the pitot tube strut for a forward sight and marking a rear sight on the cockpit windshield, we might be able to do business. The pitot tube is a pencil-shaped gadget projecting from the fuselage out ahead of the cockpit on the left, forming a V with the body of the plane. Since the pilot always looks to his left, he being on that side of the cockpit, he could concentrate on this V without having to shift his line of vision during the critical seconds of the bomb run. It seemed worth trying.

We placed an observer on the catwalk to watch the bomb drops, and I marked a tentative rear sight on the windshield with a strip of masking tape. Again and again we ran over the target at 200 miles an hour, 200 feet altitude, dropping 100-pounders without fuses. After each drop the observer called out his report and I corrected the strip of tape on the glass.

By the time we had dropped ten bombs it was obvious that the device would work, and work consistently. After a few more tries, I found that I could plant a bomb on the target with gratifying accuracy, provided the altitude and speed remained constant. Then I sat on the catwalk in the observer's position while Johnny Herron tried it, and it worked for Johnny, too. The mark we finally scratched on the windshield was just two and five-eighths inches from the bottom of the glass.

As I have said, there was nothing secret about .t other than the predetermined speed and altitude, which are no longer secret because the boys now flying Liberators in the Pacific have made changes to suit themselves, if, indeed, they use the device at all. We simply made our runs at 200 feet, with 200 MPH indicated, lined up the target in the sights, and pickled off the bombs. It was superbly simple and remarkably effective.

In attacking a ship, I usually dropped four bombs in a train, one short, one at the waterline, one aboard, and one over. Dropped from 200 feet at 200 miles an hour, a heavy bomb ordinarily will not ricochet, but the No. 1 bomb would act as a torpedo and explode against the ship's side under water. No. 2 and No. 3 would penetrate and explode on the target itself. No. 4, purposely dropped late to allow some latitude of error in range, would hit the upper works if the ship were large.

Low-level bombing is deadly—there can be no argument on that point. It is like putting the muzzle of a pistol four inches from a victim's chest and squeezing the trigger, instead of sniping at him from a thousand feet away as in high-altitude bombing. On the debit side is the fact that the attacking plane must fly practically down the target's throat, for a bomb dropped at 200 feet, 200 miles an hour, must travel about eleven hundred feet to reach the mark and will strike the target almost horizontally. The attacker, therefore, should be squared away on his bombing run at least half a mile from the target, and from there on in must travel a straight course, in level flight. Any radical maneuvering during the approach may spoil the accuracy of the drop. A well-armed target, needless to say, can make that last half mile a hot little tunnel through hell—as we were to learn the hard way before our tour of duty ended.

During February the squadron established some sort of record, flying two hundred thousand miles of search and reconnaissance, and covering millions of square miles of ocean, to assure the fleet in its Marshalls campaign that no enemy shipping was within striking distance. The fleet needed this protection, for our invasion forces had by-passed a number of powerful Jap bases. Part of our job was to keep an eye on these by-passed islands.

Wotje, some two hundred miles to the east of Kwajalein, was the worst of them. Our assignment there was to reconnoitre the atoll at frequent intervals and obtain low-level photographs revealing the extent of the installations.

It was a difficult and costly assignment. During one five-day period, two Liberators were severely shot up, a B-25 was lost, and one of our planes failed to return. Pilot of the missing plane was the same Johnny Herron who had helped me work out the "Miller Bombsight."

To obtain good low-level photographs a pilot must fly very near the target, with little defensive jinking. At Wotje, where the Jap gunners were deadly, this was akin to suicide, and Johnny knew when he took off that the mission was perilous. An easy-going, chunky fellow who looked a bit like a good-natured bulldog, Johnny for once in his life was not clowning when he climbed into the plane. The plane, incidentally, was not his own, which was laid up for repairs. He borrowed Lieutenant Commander Bundy's Liberator.

Next day, Radio Tokyo reported that a B-24 (PB4Y) had been shot down over Wotje, and Tokyo Rose triumphantly read off the names of the crew. The names she read were those of Bundy's crew, probably obtained from the plane's discrepancy log, in which his plane captain, Combs, had jotted them down to kill time on the way home from a mission. We believed, therefore, that Johnny Herron had reached his objective, that his plane had crashed or been shot down on land or in shallow water, and that some of the crew, at least, might be alive. That was all we knew and all we ever learned, despite days of search.

Two days later Lieutenant Commander Jackson Grayson failed to return from a mission over the same enemy base, and was three hours overdue before the squadron received word that he had landed at Roi, in Kwajalein Atoll, his plane riddled by AA fire, one engine shot out, and three crew members wounded. "We thought we had them by surprise," Grayson reported glumly on his return. "They're smart, those monkeys. They held their fire until we were almost on top of them, and then opened up with a barrage that almost blew us apart. The whole damned island was shooting at us." We were paying a high price for those Wotje photographs.

But the Japs were not always so clever. We caught them by surprise at Ujelang, way up near Eniwetok, after flying 975 miles to reach the target. And in mid-February, on a trip to Utirik, old *Thunder Mug* had a heckling good time at all the Jap islands en route, teasing their gunners, causing them to waste ammunition as we feinted them into revealing their positions. The Japs that day were as awkward as drunks swinging at shadows. As they swung, we calmly took notes on the location of their guns. The notes were valuable when we returned later to hit those islands in earnest.

At Utirik we bombed and set afire their warehouses and shot up other buildings during an hour's visit, then looked in on Likiep and, with one bomb left, dropped it on Wotje in the light of a low, pale moon. *Thunder Mug* never brought home bombs. In fact, with the racks fully loaded I used to store extra 100-pounders on or under the flight deck, to be hung on the shackles when the first load was expended. And the crew kept putting aboard more and more ammunition—"You never know when you'll need it, where *Thunder Mug* goes!" they said—until one day I had trouble getting the plane off the ground and discovered that Gariel, in the tail, had built up his ammunition box to hold almost more than we could fly with! "Well," the boys retorted when I lectured them, "you never told us about the extra bombs on the flight deck, so why should *we* tell all we know?"

Our reconnaissance was uncovering a lot of Jap activity at Kusaie about this time, and on the night of February 25 I led four planes in to investigate a report that a cargo ship was lying in Lele Harbor. Kusaie lies in no special group of islands. Eight miles in diameter and circled by a reef, it lifts its ragged peaks in mid-Pacific solitude, many miles northwest of Apamama. For the Japs it could have been a strong defensive outpost, menacing our sea lanes, if we had let them build it up without molestation. We were determined not to let them.

Shipping in Lele Harbor is hard to get at, however. The hills rise abruptly at the entrance, and the harbor area, sandwiched between the mainland slopes and the rugged contours of Lele Island, is so small that masthead-height approaches are hazardous. We were faced with the prospect of flying into a cul-de-sac, with defensive guns cunningly hidden on the high wooded peaks on both sides.

By scraping these peaks with our wing-tips and racing in over the harbor in single file, we not only got our ship that night but crippled many of the enemy's gun emplacements. For one of the few times in my life I lingered to watch a ship die, which it did spectacularly, exploding again and again until at last it went down with only its masts visible above a mass of floating debris. It was like watching a man writhe after you have struck him—a satisfying but not exactly a pleasant sight.

After that we hit Kusaie often, from Apamama and later from Kwajalein. But always it was a hard nut to crack (see map, p. 119) and the hidden 20-millimeters at the entrance gave us many a bad moment. Time and again we tried to plaster the worst offender with bombs, but he had set up shop with extreme cleverness, utilizing every advantage of the mountainous terrain, and he stayed in business despite all the bombs and .50-caliber bullets we threw at him. And once, in retaliation, he came within a whisper of knocking down the Liberator piloted by Lieutenant Commander Bundy, who got back only because he happens to be one of the best PB4Y pilots in the business.

For all this, the Japs at Kusaie were typically stupid. They had made up their minds that their little fortress island was impregnable, and after running the gauntlet at the harbor entrance we almost always caught them by surprise. I remember pouring in over the harbor one day and seeing a group of carpenters at work on the roof of a house. They stopped what they were doing and

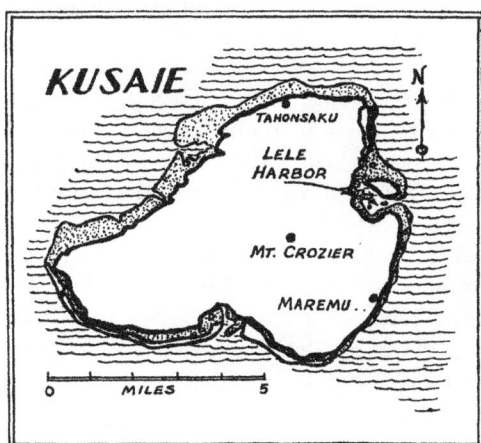

gaped in utter amazement as we thundered toward them. One, holding a board over his head, just continued to hold it there, his mouth and eyes wide open and the expression on his face plainly saying, "But you can't be American! American planes not allowed in here!"

American planes may not have been allowed in there, but American planes got there often enough, and did damage enough, to discourage the Japs from carrying out their plans. We wrecked a concrete communications building which they were patiently building. We mauled their half-finished airfield so persistently that they stopped trying to complete it. We smashed their lumber mill and burned up great piles of finished lumber with which they had planned to strengthen the island's installations.

But we never did get that 20-millimeter gun at the entrance. We just had to fly through the stuff he sent at us.

On February 25, old *Thunder Mug* earned herself the distinction of being the first four-engined plane to land on the newly captured air strip at Kwajalein, an island so devastated by our naval guns that hardly a square yard of it remained intact. Next day we were the first four-engined plane to visit the strip at nearby Majuro. And on the 28th, *Thunder Mug* went to Wake.

Wake, notorious for the vigor of its antiaircraft defenses, was a major target, never before hit in daylight by land-based Navy planes. Ours had to be a daylight mission, however, because Admiral John H. Hoover had called me out to his ship, and asked for photographs. "Good ones," he stressed. "Low-level shots. Something we can use."

We took off at night, six of us, from O'Hare Field, Apamama—Hicks, Jobe and myself of 109, Muldrow, Piper and Webster of 108, heading for Kwajalein to rendezvous with two photographic planes of VD-3 which had left Apamama earlier and staged through there. Eight strong over Kwajalein, we continued on to distant Wake.

It was a very long, hard flight, carefully planned to put us over the target in a low-level approach from the east. Section One was to fly in line abreast over Wake and Wilkes Islands, flanked by a photo plane on the port wing. Section Two was assigned a similar task over northern Wake and Peale Island, with a photo plane to starboard. Wake is roughly a sprawling V, its open end toward the west, and this plan, if we could carry it out, would take us over the point of the V and out along both arms, over every important installation on the atoll.

In bright, blinding sunlight, the sky as blue and clear as a polished globe of glass, we covered the last few miles at an altitude of only twenty-five feet, barely skimming the water. The defenders, never dreaming that their powerful base would be challenged in broad daylight, were caught quite by surprise. Most of the guns on the ground were still covered with canvas, their crews running

wildly to man them, as we flew over and let them have everything
in the book.

Very few of the Japs reached their guns, and very little enemy
fire was thrown at us until we split to traverse the arms of the V.
Thunder Mug laid four 500-pounders on the airfield; Hicks, on our
tail, leveled gun emplacements and blew up parked airplanes; Jobe
dropped sixteen bombs in a string along an area jammed with
parked planes, then blew up a fuel tank and, with a gay farewell
fillip, planted a bomb squarely on a barracks building. The gunners,
meanwhile, strafed pillboxes, personnel, planes and buildings with
such paralyzing effect that the Japs, terrified, appeared to be running blindly in circles.

As I have said before, you come home from such a raid with
only snap impressions, picked up during the brief interlude when
the plane is hurtling over the target at 200 miles an hour with its
ten machine guns yammering and bombs thundering beneath. These
impressions are vivid but disconnected. I remember seeing a group
of twenty to thirty Japanese pouring out of a shelter to man their
guns and being cut down, like tumbled dominoes, by the machine
guns in the plane. I remember the barracks building rising into the
air like a child's toy and then exploding into pieces so small that
the sky seemed filled with black rain. And I remember wondering
if the pictures would be good, and if Admiral Hoover would like
them.

With our bomb racks empty we turned for home, leaving behind
us three raging fires that filled the bright sky with pillars of smoke.
Once again we had demonstrated the value of surprise over a well-
defended target, for Wake Atoll was known to be capable of putting up a withering barrage. Yet not a plane was able to take to
the air to pursue us, and when we returned to Apamama after nineteen hours and thirty-five minutes in flight, not one of our planes
was seriously damaged.

[*For leading this audacious daylight attack on Wake, Commander Miller was awarded still another Air Medal. "The attack,"
reads the accompanying citation, "dealt a devastating blow to enemy*

shore installations, in addition to the successful accomplishment of the primary mission of diverting enemy antiaircraft fire from a photographic formation. Commander Miller's utter disregard for danger and courageous devotion to duty were in keeping with the highest traditions of the naval service."—H.B.C.]

16   *THE JOKER OF JOKAJ ROCK*

IN January we had expended much time and effort trying to destroy the Japanese airfield on Kwajalein Island. A little more than a month later the PB4Ys of the squadron were operating from that same strip.

That is the kind of war we were waging out there in the Central Pacific. As the Japs were smoked out of their island possessions and pushed back toward Tokyo, the bases relinquished by them in their retreat were used against them to extend the pressure.

The search and patrol planes of the Navy played an important role in this cross-Pacific parade. Often we were the first to investigate enemy-held islands, ranging far out from our own advance bases to heckle them, destroy their installations, obtain photographs, test their power of resistance, and sink the ships that supplied them. Then came the swift carrier task forces, to follow up our solo forays with more powerful strikes by carrier-based bombers. Thus softened, the enemy was ripe for a big-league pummeling by battleships and cruisers, and, finally, for invasion. Then the pattern of conquest would begin all over again, with Navy search planes moving into the newly acquired islands to extend their reconnaissance closer to Japan.

We moved into Kwajalein on March 7, and found to our delight that genial Lieutenant Commander Jimmy James had established a camp for us. This may sound like a very simple thing for which to be thankful, but it wasn't. Setting up a camp on Kwajalein was a man-sized job.

The entire island looked as though a super Paul Bunyan had

rampaged over it, leveling trees and buildings in a fine fury of destruction. The naval bombardment preceding the invasion had left devastation everywhere. Every tree was shredded. Every wall had been ground to rubble and dust.

American forces had been in possession of Kwajalein about a month when we arrived, yet despite days and nights of hard, dirty labor they had barely begun to clean the place up. The dead Japs had been buried, but new ones were exhumed almost daily. The enemy mines had been removed, but every now and then some poor fellow stepped on one that hadn't. For days we had to shout most of the time, to be heard above the rattle and clank of the Seabee bulldozers.

The camp which Jimmy James had set up for us was near the remains of what had been a Jap hospital. One day I explored it and bumped into a Marine who had been among the first group to clean it out. "This slab of concrete," he said, touching a piece of rubble with his foot, "was the door to an underground operating room that was sealed tight when we found it. Apparently a shell had hit close by, blocking it with debris. When we opened the room, we found nurses and doctors inside, all dead, but not a mark on them. Asphyxiated."

There was a good deal of Jap stuff lying around, even though the souvenir hunters had been prowling for days. The more I saw of it, the more I realized how difficult it is to understand these paradoxical people. They hate us with an all-consuming passion, yet many of them speak English, and some of the Japanese engineering handbooks I found were filled with English notations and abbreviations. Jap newspapers, too, were full of English words.

Often they are clever. Their Zero, for instance, is a good plane; at the start of the war it was probably a better all-round airplane than anything we possessed. Their small arms are good, especially the Nambu machine gun, which was cunningly designed to use .31-caliber ammunition—we can't use this ammunition in any of our weapons, but the Japs can use our .30-caliber in their Nambu whenever they get their hands on it, which at times has been too often. They make an airplane compass, too, which is the equal of any I have seen. Yet in the same pile of rubble that yields a brilliant piece of workmanship, you are likely to pick up something they *think* is good, but which is fit only for the scrap pile.

And even the cleverest Japs waste their talents. On many Pacific islands we found elaborate stone-and-steel houses built at what must have been tremendous effort, in localities where a thatch-roofed hut would have served the purpose just as well. We found hangars constructed like medieval temples, their steel frames fantastically spider-webbed for no apparent reason other than to satisfy the designer's whim. And the Japs are supposed to be short of steel!

Kwajalein was hot, dirty, dusty, and smelly. The inevitable coconut palms, which might have afforded some protection from the blazing sun, had been stripped or leveled by our warships' shells. Worst of all were the flies. Not mosquitoes, but flies.

Where they came from originally I don't know, but the island was swarming with them. Fat, noisy and loathsome, they attacked in droves whenever I switched on my desk light at night to do any paper work; while I worked, they walked on the paper or crawled over my arms. They were easily killed, for they had grown bloated and sluggish from feeding on Jap dead. But they disgusted us, and in spite of the mosquito netting we used, two or three of them always managed somehow to get into our tents at bedtime. Even the boys who had endured spiders and other assorted horrors in the Southwest Pacific were unable to battle the Jap flies on Kwajalein with equanimity.

But we spent most of our time in the air, now, where the flies, thank God, could not follow us. Our search sectors had been extended. Reaching beyond the old stamping grounds in the Marshalls, we ranged westward, almost to the great enemy base at Truk, exploring and photographing new harbors and shorelines.

On March 13 I decided to try for some low-level photographs of the island of Ponape, an isolated enemy fortress northwest of Kusaie. Ponape had been visited by planes of the squadron before, and we had found the Japs unusually alert there. Getting in on them in daylight without being detected during the long over-ocean approach would not be easy.

There was, incidentally, a peculiar reason for my wanting to make this mission on March 13. We were having "bogey" trouble in the squadron. Some of the more superstitious boys, remembering our record, were reminding their mates that the thirteenth day of the month was an unlucky day for us. God knows they had reason enough for thinking so. On December 13, VB-108 had lost a plane

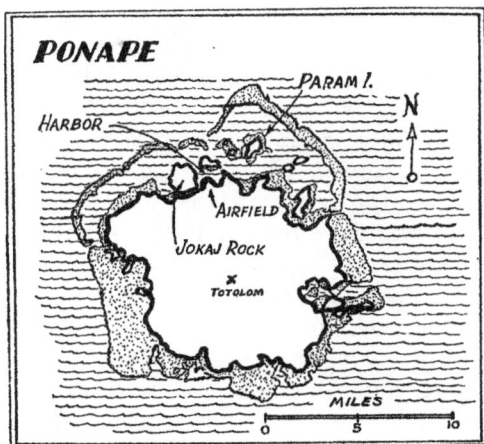

and crew. On January 13, we had lost Lieutenant Coleman. On February 13, we had lost Lieutenant Herron. But to declare a holiday on March 13, because of this, would have been too large an acknowledgment that luck played a major role in our chances. Lady Luck does play a part—any pilot is a fool who doesn't court her with all his wiles—but at the same time she must be kept in her place. Give her an inch and she'll take a life.

I studied photographs and maps of Ponape for some time before undertaking this mission, and took off at last with a plan calculated to use the island's towering peaks for protection. Ponape is larger than Kusaie but is otherwise much the same: Totolom Peak, 2,530 feet high, towers from the heart of it over lesser peaks which run raggedly down to the sea. About the whole jumbled mass lies the inevitable reef, embracing at the northern end a cluster of small, deep-water islands dominated by great, bare Jokaj Rock. On Jokaj the Japs had installed a double-purpose gun that was very bad medicine, protecting every sea approach to the sheltered anchorage behind Takatik Island.

We would get around this menace, I thought, by approaching the island from the southeast—the Japs would hardly expect a strike from the direction of their own distant base at Kusaie. This would keep the mountains between us and the airfield, neutralize the detecting devices, and give us at least a fighting chance of catching them unawares. Surprise was necessary because the pictures had to be good. Admiral Hoover had again stressed his need for sharp, clear photographs showing the extent of the enemy's installations.

We went in as planned. Hugging the water for the last fifty miles, old *Thunder Mug* swept up from the south, under the mountains, without alerting the enemy defenses. I picked a gap between the peaks and realized at the last minute it was the wrong gap, veered sharply to port to aim for the right one, and lifted the plane's nose just enough to take us through. We didn't miss by much. Those wooded slopes were deceptive, and the plane's wing tips barely cleared the sides of the funnel through which we climbed.

A few moments later we glided neatly down over the peninsula on which the Japs had built their fighter field, and there beneath us lay a neat little cluster of buildings that resembled nothing so much as a tiny civil airport in rural America. "For Pete's sake, Skipper," one of the boys exclaimed, "that looks just like a town field in Georgia!"

It was almost too pretty to destroy, but there in the midst of it, spoiling the illusion, was one of those typically intricate Japanese hangars. At 200 miles an hour we almost rubbed the roof off it as we made our run.

Robert Tovey, our photographer on this mission, was cucumber-cool as he shot his pictures. And so, I think, were the rest of us, when we realized that the Japs had once again been caught off guard. With Tovey on his camera, I banked the plane around and made a bombing run, while the gunners raked the field and mowed down the startled Japs who swarmed out of near-by buildings to man their weapons. A sharp bank to the right and we were over the town, thundering above the tinderbox houses, burning them as we went. Behind us the airfield was a shambles strewn with crumpled figures, the hangar blazed from a bomb hit, the town lifted tall thin fingers of smoke to the sky from blazing buildings. "All right," I told the boys. "Let's get the hell out of here."

We veered away, above an area in which some sort of construction was in progress—it looked like another airfield, a big one, perhaps a bomber strip from which the Japs hoped to hit us at Kwajalein. There was no time for an investigation. Jinking to escape the determined fire of the potent gun on Jokaj Rock, I turned for home. But as we went, one of the crew called out over the interphone, "Captain, did you see that construction down there? Looks like they're building something."

"It looked like an airfield," I replied.

Tovey, the photographer, matter-of-factly put an end to the discussion. "Everything's okay, Cap'n," he said. "I got some pictures of it, I think." Which was certainly a classic understatement, for when the pictures were developed they revealed every important detail of the new field, showing a string of partially completed underground bomb shelters and even Jap laborers working with pick and shovel on the near-by hillside! These and the other pictures were so outstanding that Admiral Hoover enthusiastically expressed his delight, and Tovey was given a letter of commendation, read before the men, as well as sharing with the rest of us a "Well done." Yet Tovey was not a professional cameraman. Like most of the other boys in 109 who manned the cameras on our missions, he was simply a crew member who had learned to take ictures well.

The Japs never did finish their bomber strip on Ponape. We never let them labor on it long enough without interruption. At least once a week we returned and worked them over. They tried many tricks to escape the destruction, even hiding their valuable machinery in the woods when their warning system announced our coming—in which case we would drop a bomb or two, depart, and then return from another direction to catch them at work again.

The airfield came to be a joke. "Captain, they aren't getting on with this job worth a damn," said a crew member one day. "Let's drop 'em some shovels and a note telling 'em to hurry up and finish it, so we can capture it."

Ponape was no pushover, though, with that gun on Jokaj Rock guarding the anchorage. On one occasion I tried to squeeze past it in a heavy squall, with hard rain pelting the windshield and the huge rock all but invisible from a few hundred feet away, when suddenly the gun opened up on us. The exploding shells bounced us about as though the plane were being slapped with giant paddles, and for a moment or two we were all over the sky. Hoping for better luck and better weather later, I jinked away and continued on patrol, returning in a couple of hours to find "Jokaj Joe" asleep at the switch. Unmolested, we photographed the new airfield, strafed it, and burned up a truck that was too slow to get off the runway.

After quitting the island on this excursion, we found a Jap ship three miles off shore, apparently on her way in with a load of cargo. She was a surprised Jap when *Thunder Mug* laid a bomb in her path. Frantically she zigzagged, leaving in her wake a train of suds that looked from above like the vapor trail of a stunting airplane. One at a time I pickled off the bombs until one of them hit home and stopped her. Then, as some of her crew struggled with a lifeboat and tumbled into it, we finished the job with machine guns.

I seldom hit Ponape without looking in on the near-by tiny islands of Ant and Pakin as well, and on this day, proceeding to Pakin, I discovered another Jap cargo vessel, her decks so laden with stores that she was too sluggish even to maneuver. As we made our attack at masthead height, with all guns working, her crew, too, struggled to launch a lifeboat, and one man knelt behind the deckhouse, terrified, with his hands clapped over his face. An instant later the deckhouse itself burst apart like a cardboard firecracker, and through one shattered window came a man's coat—we thought at first it was a man—flapping weirdly in the wind. Then with her deck ablaze and her cargo sending forth billows of smoke, the ship turned crookedly toward the reef. There she died. When we left her after six strafing runs, she was fiery red from bow to stern, with all her ribs laid bare.

Still *Thunder Mug's* day was not finished. From Pakin we flew on to the little island of Nikalap, dropping incendiary clusters on a row of houses near an unfinished installation that looked something like an airfield apron—though Nikalap had no airfield. Whatever it was meant to be, the Japs never finished it. The wind was favorable that day, and when we left, every last one of their buildings was blazing beyond control. Then, not reluctantly, we headed for home.

[*Commander Miller was awarded his first Distinguished Flying Cross for this mission, the citation reading as follows: "For heroism as Commanding Officer of Bombing Squadron 109 in action against enemy Japanese forces in the Eastern Caroline Islands on March 25, 1944. While approaching the Island of Ponape on a regular search mission, Commander Miller photographed and effectively strafed hostile installations, then sighted a Japanese coastal vessel off shore, daringly attacked at masthead level, causing it to blow up and sink immediately. Proceeding to Pakin Island, he strafed another ship, forced it into a reef and left it burning from bow to stern, and, later, flying over Nikalap Island, set fire to all buildings with incendiary clusters. By his great personal valor and outstanding flying skill, Commander Miller contributed to the ultimate defeat of the enemy in this area."—H.B.C.*]

For several weeks the squadron found good hunting on these forays to Ponape, despite the ambitions of our friend on Jokaj Rock. And there were lighter moments, too. One day in *Thunder Mug* we came across a small sailboat, a mile or so from shore, in which a Jap and his lady-friend were apparently just out for a ride. It looked for all the world like a Sunday morning off Cape Cod.

"They don't know there's a war on," said Delahoussaye.

We could have sunk them with a burst of machine-gun fire, and one of the gunners, I remember, wanted to. He thought it might be a trick of some sort. But I vetoed the suggestion and we flew on, leaving the Jap and his lady to their blissful contemplation of nature.

On another occasion we went into Ponape at night to get the gasoline dump, and, for luck, finished off the hangar, too. They had spent a lot of time on that hangar, repairing it every time we bombed it. This time it burned to high heaven. When we last looked back, its intricate steel framework was cherry red against the night, a Coney Island roller coaster lit with neon.

Meanwhile, the Japs were beaver-busy on Ant Island, building boats for their interisland commerce. To protect the place they had installed radio equipment in a towerlike structure that resembled a windmill stripped of its fan. A surprise attack wiped out their windmill. Incendiary bombs put an end to their boat-building. The Japs were being forced to abandon a good many of their pet projects on these islands, and no doubt it annoyed them.

On April 9, Easter Sunday, we decided to visit Ponape for a special Sunrise Service of our own, and took along thirty 100-pound bombs on which the boys had splashed some bright paint. "Easter eggs," Gariel remarked dryly. "We'll have an egg-rolling on the Japs' front lawn." Arriving at sunrise, we rolled our eggs as planned, wiping out at least a week of hard work which the Japs had spent on their airfield. But as we turned away, the joker on Jokaj Rock suddenly found our range.

"Captain, No. 3 engine is smoking," one of my mechs said anxiously.

He was right, but he didn't know the half of it! One of Jokaj Joe's 20-millimeter shells had mangled the cockpit, and the controls were gone on the throttle, cowl flaps and supercharger. I suddenly realized how very far we were from home.

"Smoking?" I said breezily. "Hell, that engine's too young to be smoking." Corny, yes. But there are times when a shot of pure corn is more beneficial than medicinal whisky. "When you smile,

Skipper, everything is all right . . ." So I grinned and cracked a corny joke, and the moment of tension was past.

"We'll go home," I said then, "instead of continuing the patrol." But as we left Jokaj Joe behind, the engine stopped smoking and I saw beneath us two fine Jap ships tied up beside the Ponape lumber mill. They were too good to miss, and so we turned to get them, made twelve passes over them and burned them at the dock, the boys at the guns blazing away with their usual efficiency just as though we didn't have a No. 3 engine that might poop out or catch fire again at any moment. We could have got home on three engines anyway.

As for the 20-millimeter shell with which Jokaj Joe saluted us, I found it when we got back, and still have it. It is a good shell, not even scarred by its passage through the metal of the plane.

So it went, as we did our best during those March and April weeks to keep Ponape impotent despite the Japs' attempts to build it into a fortress. At one time or another most of the planes of the squadron paid it a visit, and the reports usually read something like this:

"On routine search, destroyed cargo vessel off shore, started fires on barracks and hangar, dropped incendiaries and strafed radio station and town area, strafed new airfield causing personnel casualties; no damage to plane or personnel, pictures taken."

Except that, like *Thunder Mug*, some of the planes returned with souvenir holes after tangling with that very good Jap battery on Jokaj Rock, which we never did knock out, and which the Army later tried to knock out with equal lack of success.

The pay-off? On the first day of May, the new, big warships of Vice Admiral Marc Mitscher's Pacific task force moved in on Ponape to give the Japanese occupants a final dose of discouragement. The ships' gunners had difficult targets, many of them cleverly hidden on the heavily wooded slopes of those rugged Ponape peaks. But they knew where the targets were, and they hit them.

VB-109 was not there that day. We were looking in on other Jap-held islands closer to the enemy's homeland. But we were present in spirit, for along the bottom of many a photograph studied by Admiral Mitscher's men in preparation for the bombardment was the formal Navy notation that VB-109 had done the photographing.

17   *YOU SPELL IT T-R-U-K*

ONE day on Kwajalein I wrote a letter to the minister of the church back home in Winston-Salem.

Letter writing is not a habit of mine. I frequently don't get around to those that really ought to be written. But life on Kwajalein was primitive. We lived in a sector in which the Japs had maintained a communications station; you had to pick your way through old Jap receiving sets and blasted telephone equipment to get to the head. The smell of dead Japs still hung over that part of the island, and the air was so full of dirt and dust that we had to shake out our beds at night before getting into them.

Kwajalein made no concessions to the niceties of living. It was simply a base from which to carry on the business of killing. Yet when I walked down to the enlisted men's tent area one day I found that despite the press of war, someone had found the time and the inclination to nail a sign to a palm stump at the entrance. On it were stenciled the words: "GOD'S COUNTRY. ANYONE WELCOME!" And so I wrote to the minister back home, thinking he would be as impressed as I was.

That sign was good for me, for in addition to the miseries of life on Kwajalein, these were days of long and hard patrols and our maintenance headaches were increasing. Pilots frequently took off two or three times, in different planes, before finding a plane that would complete the mission. Men and planes were tired, and there had been a tendency to forget that, after all, we were human beings and not just mechanical parts of an airplane. The sign reminded me that old Shafe, my navigator, had a grand sense of humor; that Paul Ramsey, the plane captain, still wrote letters to his WAVE in Jacksonville; that Gariel, Dorris, Johnson, Delahoussaye, Jaskiewicz, Simmen, Young and the rest of the boys had been a lot of other things before necessity turned them into Jap hunters. And that I owed Thelma and the kids some letters.

Lieutenant Commander Jimmy James, the Aircraft Service Commander who maintained our planes, had this feeling too, I'm sure. In the center of the camp he had set up a big rustic stump, neatly peeled, on which he encouraged the boys to carve their initials. "Some day," Jimmy told them, "you may want to come back here as tourists. Then you can point to your name or initials on our 'Plymouth Rock' and talk about what you did here. Because what you're doing here is important."

Almost everyone in the camp had his name or initials on that stump.

From Kwajalein our principal targets were Ponape, Kusaie, and military installations on lesser islands in the Eastern Carolines. There were brushes with enemy submarines, one of which lasted night and day for seventy-seven hours, and there were occasional run-ins with Jap planes. We made reconnaissance flights to Wake, on one of which Lieutenant Elmer Kasperson got what he thought was a submarine and discovered a top-hatted Japanese who could almost outrun a Liberator.

The "sub" turned out to be a small cargo ship (it looked like a conning tower above the long, level lines of the pier at which it was moored) and old Kas blew it to smithereens anyway. The top-hatted Jap was one of a group of six Wake Island occupants who were strolling down a road when Kas peeled off to do some strafing. Five of the Japs dived into a ditch. Top Hat took off down the road with his coattails flying. "And so help me," Kas said later, "we just couldn't overhaul the guy, he was traveling so fast!" Which was a bit of an exaggeration, for of course the Liberator did overhaul him. But by that time the boys were laughing so hard that they didn't have the heart to reduce the population any more, and they let Top Hat continue his marathon.

Lieutenant Hal "Bird Dog" Belew discovered something while

we were based at Kwajalein, too. He learned that a Liberator can be landed with a full load of gas, a fact which may be of some interest to the manufacturers, who say it shouldn't be done. Hal took off one night with a maximum gas load and four 1,000-pound bombs, and had barely cleared the ground when one engine quit. He dropped his unarmed bombs in the lagoon, circled the field and requested the tower to turn on the lights again. For some reason the request was ignored. So Belew came in anyway, putting his Liberator down as nicely as you please in the dark, on three engines, with all that gas aboard. (You can't dump gas from a PB4Y as you can from some planes; it has to be drained out.)

But in general there was little of the spectacular in the work we did out of Kwajalein. It was a grind. All of us flew too hard and too often. On the third day of April, for instance, the April entries in my flight log showed fifty-three hours in the air—fifty-three out of seventy-two—and that sort of thing is wearing. We needed not so much a rest as a change from the routine patrols and the raids on Ponape, Kusaie, Wake, and other islands which we knew by heart.

On April 3, the crew of *Thunder Mug* at last had a taste of something different, and I personally satisfied an ambition of many months' standing. It began with a report that an enemy carrier and four or five escorting destroyers had been sighted at Fananu Island. Look at that name again. F-a-n-a-n-u. No other details were available, and a study of the map revealed that Fananu was a little trading station on the eastern rim of Nomwin Atoll, in the Hall Islands, about seventy-five miles north of Truk. We had never been there—it was a long way from Kwajalein. But a Jap carrier was big game and worth the attempt.

Using two planes, *Thunder Mug* and Lieutenant Jadin's *Helldorado,* we departed from Kwajalein in the afternoon and staged through Eniwetok, which American troops had captured February 20. Gassing up at Eniwetok, we left there just after sunset, planning to rendezvous near the target and go in together. But it was a bad night for the attempted rendezvous. The moon was diffused by a thin, hazy overcast, visibility was poor all the way, and the two planes, unable to effect a meeting, proceeded independently to the Hall Islands. Lieutenant Jadin got there first, made his search, found nothing and returned to Eniwetok.

In *Thunder Mug* we arrived above Fananu about two o'clock in the morning, quite a while after Jadin had left. This need not be surprising. I have a reputation for being a slowpoke, and despite certain claims made by wild-eyed correspondents, I believe firmly in the old aviation adage that there are no old, bold pilots—there may be old ones and bold ones, but never old, bold ones. At any rate, we flew over Fananu and Nomwin Lagoon, hunting for the carrier in the strange, milky light of a fragmentary moon half hidden in the overcast, and, like Jadin before us, found nothing. Nothing, that is, except some fairly accurate AA fire, for Jadin had alerted Fananu and in that eerie half-light the Japs could see us better than we could see them.

With our load of four 1,000-pound bombs still intact, we turned back, cussing out the report which had sent us such a distance for nothing. Then I began getting an idea. "Shafe," I said, "maybe they spelled that name wrong. There's a Fanamu Island, too. F-a-n-a-m-u. It's in Truk. What's the course for Truk, Shafe?"

Old Shafe didn't blink an eye. He knew darned well that I'd been carrying a folder of Truk maps in the plane for weeks. He knew that I had talked long and often with Ed Lloyd about the location of the anchorages in this so-called "Japanese Pearl Harbor," and that Ed had marked the target map for me. And, of course, Shafe had known for weeks that sooner or later we were going in to Truk to find out what the Japs really had in there.

"All right, dammit," he said, reaching for my Truk folder. "I'll go anywhere you will."

The co-pilot said, "Captain, this is crazy. No plane ever went in there alone before." Which was wrong. What he meant was that no plane had ever challenged Truk alone at masthead height with the idea of doing damage. A Marine photographic plane had been in at very high altitude to take pictures, and Army planes had been in high, mostly at night, to drop bombs. I mentioned these facts, and also that we had four 1,000-pound bombs to get rid of. "No sense in taking them back to Eniwetok," I insisted. But he had heard a lot about Truk—we all had—and didn't like it. At which point I could have gone philosophic, expounding the theory that we'd all be a lot more scared the next time if we turned back now, but I let it pass. Old Shafe was already figuring out our course.

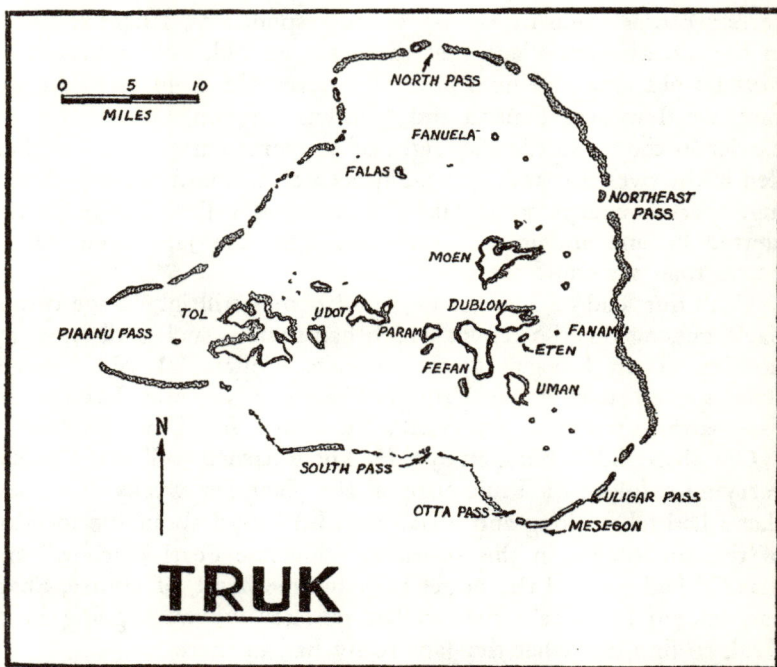

Truk, reported to be the strongest enemy base in the Pacific, capable of holding the entire Japanese navy, is a forty-mile-wide lagoon containing half a dozen major islands and scores of minor ones. Its harbor facilities are among the best in the Pacific, protected by well developed air strips on both major and minor islands. It is listed in naval textbooks as one of the best natural fortresses in the world.

All right. But most of the islands are mountainous, and we had become fairly adept by this time at using mountains for a shield. Moreover, the Japs certainly would not expect us. On the books our little expedition would undoubtedly be called "audacious" and "daring" (It was!) but in my mind I was pretty sure we could get in, do some important damage, and get out again without running any more than the usual risks. The enemy would be more apt to expect carrier planes (Vice Admiral Marc Mitscher's carrier planes had hit them in February) or high-flying Army bombers. They wouldn't be looking for a single low-flying Liberator.

Over the interphone I told the crew where we were going, and less than half an hour later, still in that pale, milky mist of moonlight, we crossed the Truk reef at 100 feet altitude, just east of North Pass. Below, the coral shallows of the reef and the breaking waves were luminous with the natural phosphorus of the water, and winked at us.

Our target, selected from a study of the map on the run down from the Hall Islands, was the main anchorage west of the major island of Moen, in the heart of the Truk defenses. (See map, p. 138.) With luck we would find some warships there. On the way down we skirted Fanuela Island, keeping our distance lest they spot us and give the alarm. The island was a dark blot against the pale water, and a second blot, a Jap destroyer, lay a little distance to the south. "We'll get that," I told the boys, "if nothing better turns up." At which point one of the crew, noting two dim red lights on the island or near-by shoals—obviously navigation lights of some sort—called up through the interphone, "Look, Cap'n, they even turned the lights on for us!" How can you miss with a crew like that?

Still undetected we continued south, low over the water with the mountainous Truk islands looming all around us, until we had traveled the whole length of the main anchorage and were off the northern tip of Fefan. The anchorage was a disappointment. It contained a single cargo ship of five or six thousand tons riding blissfully at anchor. No warships. And now, of course, we were tense, for we were deep inside the defensive heart of Truk, and the islands hemming us in were known to be bristling with potent guns, and at any moment the Japs might pick us up in their de-

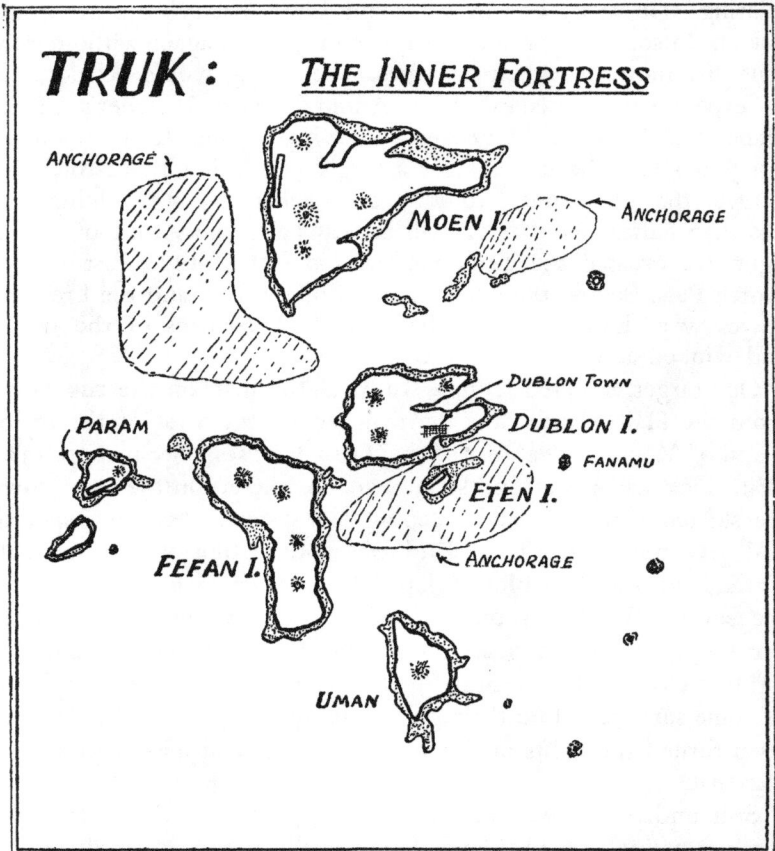

TRUK :   THE INNER FORTRESS

ANCHORAGE

MOEN I.   ANCHORAGE

DUBLON TOWN

PARAM   DUBLON I.

FANAMU

ETEN I.

FEFAN I.   ANCHORAGE

UMAN

tecting devices. No one spoke. We flew almost on the water, feeling our way through the dark. Not a shot had been fired.

There *had* to be some Jap warships in here somewhere, I told myself, and so, circling to port, I flew northward again, this time hugging the western shore of Moen Island. In circling, I took a long look at the pass between Dublon and Fefan—what a chance to surprise anything the Japs had hidden there!—but I had faith in the map Ed Lloyd had marked for me, showing the location of the main anchorage, and resisted the temptation to go exploring. Perhaps there would be enemy ships tucked close inshore against Moen. But there weren't.

Disappointed, we swung back to search once more the area between Moen and Dublon—Ed had marked a smaller anchorage off the southeast shore of Dublon. But now we had overplayed our luck. The Japs, suspicious at last, went for their searchlights, and inquisitive beams of white light began reaching up from the islands. One or two of them licked over us, then frantically tried to find us again as we came about to head for what obviously was the most worthwhile target: the destroyer up near Fanuela Island. Tracers climbed up the searchlight beams, hunting for us. As we ran northward, the area between Dublon and Fefan suddenly became hotter than a flame-thrower.

We almost missed the destroyer. The dim, diffused light of the moon through that milky overcast was deceptive, disclosing mountainous land masses at a distance of five miles or more but muddling my judgment at close range. I was on top of the ship before I saw her—she suddenly materialized in the mist only half a mile ahead. Failing to pick her up in time, I had to come around for a second run, and then a third, before I had one to my liking. By then, the shooting and the searchlights on Moen, Dublon and Fefan had alerted the ship's crew, and they were on their guns.

There was time for just one bombing run as the destroyer's AA fire came up like red rain to meet us and the searchlights on the northwest corner of Moen, locating us at last, pinned us in the glare of two crossed beams. Old *Thunder Mug* must have looked like a gigantic moth in a candle flame to the dark little shapes rushing about on the deck of the Jap ship. Low and fast, we screamed over them, dumping our four 1,000-pounders.

The bombs stopped the destroyer fire. They went off almost in unison, producing one hellacious explosion that broke the ship in two and sent her stern end swirling away from the rest of her. But the gunners on Moen had us in their sights then, and the crossed searchlights stuck like glue. And we were so blinded by lights from other islands, weaving wildly over the sky, that the co-pilot reached for his polarized glasses.

For thirty seconds or so, old *Thunder Mug* all but shed her skin as we jinked to get out of the lights and the AA fire.* We were all over the sky, weaving, diving, climbing—and all over the water, too, with the belly turret skimming the surface. What finally cleared us was a steep dive toward the water, a feint to port, a fast climbing turn to starboard which carried us to the outer edge of the circle of light. Before they could center us again we were in darkness, "getting the hell out of there" at a speed that made the wind sing.

Behind us the searchlights angrily swept the sky in vain, and the AA gunners on Moen went right on shooting at where they thought we ought to be. The destroyer, burning, was done for. The Jap gunners were still shooting at shadows when we crossed the reef on our way out of the enemy's "impregnable fortress." They hadn't hit us—much—but if it is any consolation to them, they did scare the living daylights out of us. I was still sitting in a puddle of sweat when we arrived at Eniwetok at 9 A.M.

We had sent in a flash report, of course, after leaving Truk. At Eniwetok, Captain E. A. "Bat" Cruise, the Island Commander, was waiting for us. Captain Cruise is an old-time aviation man, well loved. I had breakfast with him. But there was no time to clean up or rest; they were waiting for us at Kwajalein, too. And so at two o'clock in the afternoon, after being out nearly twenty-four hours, we arrived back on the strip at Kwajalein, all of us dead tired, dirty, and of course still wearing the clothes in which we had sweated out our pilgrimage through Truk. And there, waiting for me in Cap-

* *Comment by Lieutenant Edward W. Lloyd, USNR, ACIO of VB-109: "The maneuvering of a PB4Y plane at these low altitudes at night calls for a high degree of skill and daring, for the pilot, while searching for his target and then dropping his bombs in the face of AA fire, must be constantly engrossed in keeping his plane from striking the water. The accuracy of the bombing obtained by such methods, however, is ample justification for the years of practise and experience required to successfully deliver such an attack."—H.B.C.*

tain Taff's tent, was our administrative boss, the Commander of Fleet Air Wing Two, Admiral John Dale Price!

The admiral didn't mind the dirt, sweat and whiskers, however. He was a very swell guy. After I had told him about the flight, he informed me that Admiral Hoover had received our flash report, too, and wanted me out on his flagship. "Go on out there," Admiral Price said. "Never mind the beard and the shirt. Aviators are supposed to look like that after being out twenty-four hours." But some kind soul had telephoned the ship, and a message came back from Admiral Hoover advising me to sleep first and see him later.

As a matter of fact, I didn't sleep. I had reached the wound-up condition where sleep is impossible. So, after shaving and bathing, I began to pack my things. For this was our last day on the island. Next day the squadron moved to Eniwetok, advancing another step along the road to Tokyo.

[*From an Eniwetok Island bulletin, 4 April, 1944: "Commander N. M. 'Bus' Miller, Commanding Officer of our new Navy Liberator Squadron, took on Truk last night single-handed. He and his crew took a Liberator out to an atoll near Truk to blast a reported Jap carrier. He searched the atoll thoroughly and, finding no carrier, decided there was no use hauling all those bombs back here, and since the nearest good target was Truk, he went to Truk. Commander Miller came in very low, and the Japs lit navigation lights for him on the reef. They probably would have turned on the airfield lights and produced sake (Japanese wine) as well, but when the Liberator opened her bomb bays and dropped some 1,000-pounders on a Jap destroyer, they got sore and started shooting. Then it was too late. The PB4Y got in this morning."—H.B.C.*]

18   *SIX OF EIGHT*

FROM the air, Eniwetok Atoll looks like a necklace laid out on a jeweler's counter, with the clasp, Eniwetok Island, at the bottom.

The island itself is crescent shaped, two miles long and not much more than a third of a mile wide, and is split lengthwise by a solitary road running to the airfield.

Like Kwajalein and Tarawa, Eniwetok had been blasted almost into oblivion by the big guns of our warships. Hardly a building or a tree remained standing when the Marines went ashore to take the place over. Yet when Squadron 109 arrived, six weeks later, the island was fast becoming a very livable spot. The dead Japs had been buried, mines and duds had been removed, cargo ships were steaming in and out on schedule, and Seabees of the 110th Battalion were looking around for new jobs to test their talents.

There are "happy" islands, just as there are "happy" ships. Eniwetok was one of them.

"We have a remarkably healthy place here," the doctor told me. "The days are not too hot, the nights are pleasantly cool. No colds here. Just a little malaria and jaundice, brought up from the Russells by some of the Marines. No trouble with flies or mosquitoes, either."

Remembering the Jap flies on Kwajalein, I hoped he was right. And he was. A steady wind kept the island remarkably free of insect pests. The wind was a blessing in another way, too. Before long the boys had rigged up little windmill washing-machines outside their tents, using discarded crankshafts from the island junk pile. In the morning, before going out on patrol, they would toss in their soiled clothes. When they returned at night, the washing was done!

These were busy days for the squadron. On April 18, after participating in a mission to the outer fringe of Truk Atoll, I circled south to look in on Mesegon Island, where the Japs were reported to be building an air strip. The strip was there, but was still dotted with tree stumps, and we decided to let the Japs do a little more work on it before discouraging them. Before departing, however, we explored the southeast corner of the Truk fortress and in the darkness located a 2,000-ton cargo ship. The night was ink black, and we probably would not have spotted the ship at all except that our approach was at masthead height and the vessel's superstructure was outlined murkily against the lesser darkness of the sky.

We attacked, and as the boys cut loose with their machine guns I let go a stick of bombs. One was a direct hit. When we left the lagoon a moment later a column of black smoke, red with tongues of flame, was curling up to blend with the darkness above another dying Jap.

[*Commander Miller is being too matter-of-fact again. This attack won for him his Fifth Air Medal, not so much for sinking another enemy ship, but for the aggressiveness that prompted him to go on the prowl for another ship to sink. The citation reads, in part: "After successfully completing an extremely hazardous reconnaissance mission in company with other planes near the entrance passage of Truk Atoll, he left the other planes and continued on alone at low level into Truk Lagoon. Sighting an enemy cargo ship half way between Uman and South Pass, he strafed it at masthead height and obtained a direct bomb hit. Black smoke poured from the stricken ship and it was either sunk or heavily damaged. His actions and devotion to duty throughout were in keeping with the highest tradition of the U. S. Naval Service."—H.B.C.*]

A few days after this second solo sortie to Truk, we had a time for ourselves on a "routine search mission" and incidentally provided artist Joe Domarecki, whose paintings adorn so many officers' clubs, with some fairly dramatic material.

Admiral Price had sent Ensign Domarecki out to Eniwetok to put some of the Navy's Central Pacific activities on canvas. He had done the island thoroughly and was eager for new material. Learning that *Thunder Mug* was scheduled for a flight this day, he showed up at my tent and asked to be taken along.

"We'll be out all day, Joe," I warned him.

"All right," he replied, grinning. "Just keep out of Truk. I'll paint Truk after we capture it, if you don't mind."

At 6:30 in the morning, we took off and began what was supposed to be nothing more than a routine milk run extending through the western Carolines as far as Namonuito Atoll, a little cluster of islands lying north and west of Truk. It was a beautifully clear day, and I took advantage of this to give the islands a thorough looking over. Each of them had its small group of native houses. Each, in addition, boasted at least one Jap-looking, thatch-roofed hut which gave evidence of being a lookout post for the island.

On Ulul, one of the largest in the group, there was no doubt whatever as to the purpose of a cluster of V-roofed structures surrounding a radio tower; it was a typical Jap radio and weather station. And so we burned it to the ground with incendiaries, stirring up quite a conflagration. Smoke from the funeral pyre rose serenely to the heavens behind us as we turned away.

"Wonderful," Domarecki said, patting his camera. "I've got that in the box here, for the record."

I thought he would be disappointed when he saw the pictures, for these enemy outpost stations, if placed on a New England hillside, would have been no more impressive than farm buildings slightly gone to seed. Yet to the Japs they were immensely important as part of the network of communications by which the enemy hoped to keep a close check on our Navy's movements. Any time we could destroy a link in that intricate chain, we were making matters easier for our task forces.

Namonuito being the western terminal of our search sector for the day, I came about over Ulul's burning buildings and made for the two atolls of the Hall group, passing Fayu, which had nothing to offer. At Igup, in the westernmost atoll, the Japs had completed a large building, probably a storehouse of some sort, and a cabin boat, partially hidden, was drawn up on a near-by dock. I swept in low, and while I dropped an incendiary on the boat, the boys worked the buildings over with their machine guns. Boat and buildings were both ablaze when we set sail across the ten-mile stretch of sea to the eastern atoll. Domarecki was feverishly snapping pictures at every opportunity.

Expecting the Japs at Ruo Island to be on the alert for us, I approached low on the water. They saw us coming. As the plane climbed to clear the palm trees, five coastal vessels in the lagoon beyond began scurrying about like water-bugs on a mill pond. Five of them! I glanced at Domarecki, and his eyes were as bright as the AA fire the Japs were tossing up to screen themselves.

During the following fifteen minutes, *Thunder Mug* covered just about all the sky there was above those five zigzagging ships, in as weird a game of tag as the Japs had probably ever played. I dropped bombs, scoring a number of near misses, but most of the damage was done by the boys on the machine guns.

These were wooden ships. The machine-gun bullets spattered around and over them, leaving little rivers of fire that ran along their decks and blossomed into explosions. Such fires, once started, were uncontrollable, though as we jinked over the vessels again and again, piling ten or twelve strafing runs on top of the initial bombing attacks, we saw the Jap seamen struggling to fight the flames with anything on which they could lay their hands.

Ships and shore were after us then with AA fire, some of it heavy and some of it fairly accurate, but we had hit them too suddenly and sapped too much of their strength for them to be effective. Hit but not hurt, we continued the attacks until convinced that all five ships were doomed. One had exploded again and again, pumping out gusts of yellow smoke. Two were ablaze to the waterline, gutted beyond repair, and the remaining two were vainly struggling to make way through spreading pools of oil. All over the lagoon lay parts of ships torn loose by machine-gun fire and bomb explosions. So, with a few retaliatory bursts at the shore installations, we departed.

This time Domarecki wore an expression of pure rapture as he hugged his camera. "Man, what pictures!" he exclaimed.

I looked at the gas gauge. We had been out ten or eleven hours and were low on fuel. The bombs were gone from the racks. But this obviously was our lucky day, and the boys eagerly hung up some extra bombs which I had lugged along under the flight deck. Murilo Island at the eastern end of the lagoon was only eight miles distant, and if the Japs were sending ships to the atoll on a wholesale scale, we just might find more of them there.

We just did. Sitting pretty in the shelter of the island were two enemy coastals and a patrol vessel, gayly bedecked with flags.

Maybe this was the Emperor's birthday! They saw us coming—probably they had been alerted from Ruo—and greeted us with a hot barrage of AA fire, at the same time hauling down the flags and maneuvering around the lagoon to get out of our path. The two coastals were big game, larger than the five ships we had left burning at Ruo. Probably, with the flags, they were naval auxiliaries. They maneuvered well; our remaining two bombs were wide. But we still had machine-gun ammunition, and *Thunder Mug's* gunners were hot. In fact, one of them was too hot. Whisky Jazz Jaskiewicz, pounding away with his port waist gun, suddenly found himself flat on his back with the gun in his lap. It had broken loose from the mount and knocked him to the deck, and before the kid from Chicago could relax his grip and cease firing, he had shot some sizeable holes in the fuselage. Fortunately he suffered only a bruised back, and the starboard waist gunner, Delahoussaye, helped him to his feet as I came about to continue our attack.

As the ships circled, I circled with them, heading them off, turning them back. Half a dozen times we changed their minds for them with machine-gun fire, until suddenly they abandoned any hope of escape. The patrol vessel and one coastal ran for the reef to beach themselves, staving in their bows as they plunged into the coral. The second coastal burst into flames a mile from shore, shuddered under violent explosions, stopped dead in the water, and obviously would require no further attention.

Leisurely, then, we finished off the two on the reef, strafing them as their crews swarmed over the sides into the water to swim the two hundred yards to shore. Both were hopelessly afire and the surrounding water was littered with debris when we quit. By then our guns were hot, and only enough ammunition was left for protection on the way home.

As we pulled up and away, I turned to Domarecki. "Did you get pictures of that?" I asked.

Glumly he shook his head.

"You didn't? Why not?"

"I used up all my film before," Joe said sheepishly. "I didn't know you guys stayed out so long."

Well, it had been a long time, and when we finally got back to the base we had been fifteen hours in the air. I knew how our passenger felt. My own duff was full of aches, too. But the trip had been worth it. When one of the squadron planes went out to obtain pictures the following day, those Jap ships—what remained of them—were still where we had left them. At Ruo, where we had caught the first five, two were sunk near the beach, two were sunk farther out in the lagoon, and one, in deeper water, was submerged. At Murilo, one of the coastals had gone down in deep water and the patrol vessel had drifted across the lagoon to the opposite side of the reef, where she lay beached and deserted. So, for the record,* we had destroyed at least four of the five and two of the three, for a six out of eight total. The other two were marked down as damaged. And Ensign Joe Domarecki, despite the lack of photographs, painted the attack scenes from memory, and they were good.

[*From Admiral Hoover:* "After hearing details, I consider sinking and damaging 8 vessels by search plane of VB-109 to have been an outstanding and remarkable performance throughout. Extend my heartiest congratulations to plane commander and crew."

From Eniwetok Island bulletin, 23 April, 1944: "Out in the far wastes of this ocean a lot of people from here are, in the daily routine of their business, being what newspapers call 'heroes.' It's hard to spot them around here. One of them is a big, leaned-out, rugged North Carolinian whose commander's leaves are green with verdigris and whose cap has spent so much time jammed into some small space on a Liberator that it is undoubtedly the most beat-up affair in the Navy. Commander Miller, the one-man war on Truk

* Bus Miller was not too concerned about the "record." From Commander J. E. "Bucky" Lee, USN, comes a copy of a letter Bus wrote in reply to a request for his final estimate of ships sunk and damaged on the above mission. The requested information is supplied with conscientious attention to detail, but Miller's letter concludes with the following postscript: "Isn't this a helluva lot of gum-beating about nothing? I've been hanging around these ACIOs (Air Combat Intelligence Officers) too long!" Incidentally, this was the mission for which Commander Miller was awarded his Second Distinguished Flying Cross.—H.B.C.

*a little while ago, is sweating out no medals and where he goes the
grandstand is occupied by Japanese.*

*"A few days ago Miller took his Liberator (which hasn't even
got a naked woman painted on it) out on routine search and reconnaissance. He searched and reconned as much as required, but had
no luck. Then—like these shows where the girl comes on in a long
crinoline gown and waltzes around to corny music for a while, and
then when the band breaks loose and goes out of the world the
crinoline comes off (leaving the girl in two airmail stamps and a
transfer pay account) and she goes into a dance which would raise
your hat—like that, Miller and his airplane gave up sedate sailing
around the sky and got down to business.*

*"He flew to Ulul Island first and blew up the only Jap building
on the place. He did the same for the Japs on Igup. Warmed up, he
went over to Ruo where he found 5 small transports. (These 'small
transports' are the very blood of Japan in the islands—when we get
them all, the Japs can whistle for food and ammunition.) He sank
two and damaged three, and when he 'damages' something it is very
damaged. Then he went over to Murilo where he found two more
small transports tied up to a patrol ship. He sank one transport and,
being fresh out of bombs, chased the other two ships around the
bay with his guns until both of them, 'damaged,' went aground.*

*"This one man task force goes around Eniwetok just like ordi-
nary people. When he finishes washing his shirts he doesn't even put
his theatre ribbons on."—H.B.C.]*

19   *"LET'S DON'T LOUSE IT UP"*

"Jonesey," I said one day, "will you kindly leave this forsaken little
island alone, so that once in a while, when we return from a flight,
we'll be able to recognize it?"

Commander Jones grinned happily. He knew how good his Seabees were—they were amazing. We used to watch them at work,
and we simply didn't believe what we saw. Having by this time

attended to all the essential jobs on the island, they were killing time by transforming junk piles and scrap heaps into structures of outstanding beauty.

"Jonesey," we'd say, "when are you going to build the Lincoln Memorial out here?" And Jonesey would grin.

Living on the same island with Commander Jones had its drawbacks, however. At times the inhabitants of Eniwetok would gladly have taken up housekeeping on the rim of an active volcano, just for a little peace and quiet. The Seabees were everlastingly blasting for coral in the lagoon, and at all hours of the day or night, as you went about your business annoying no one, the island would suddenly shake as if seized with convulsions. Then a mighty roar filled your ears, and a fountain of water and coral fragments would hurtle skyward, some six hundred feet high, in your immediate vicinity.

Jonesey never called his shots. On one occasion a coral rock as big as a football burst like a comet through the walls of Commander Eddie Renfro's tent. That the tent was empty at the time was purely Eddie Renfro's good fortune.

Few of us complained about these minor annoyances, however. When we did, someone was sure to repeat the story of the Marine on Guadalcanal. It happened, so they say, just after the six-months campaign on that island had ended. A Marine sergeant, veteran of the initial landings, was shambling peaceably down a road, tired but happy in his beaten-up shoes and tattered uniform. Toward him strode a resplendent, newly arrived Army colonel. As they passed, the sergeant's salute was not to the colonel's liking.

Snapped the colonel: "Sergeant, this island is United States territory now! We have the usual formalities to observe!"

Replied the Marine, rubbing a trouser leg worn thin by six months of intimate contact with the jungle and the Japs: "Now look, bub, we got something real good here. Let's don't louse it up."

I heard that story on Apamama, Kwajalein, Eniwetok, and later on Saipan. By now it must be a Pacific classic. On Eniwetok, the man who got the biggest kick out of it was undoubtedly Lieutenant Colonel Thompson, commanding officer of the 10th Defense Battalion. Colonel Thompson was a veteran Marine Corps man, proud of his outfit. His battery of 90-millimeter guns in the Russell Islands had set some kind of record during one of the Jap raids by shooting down fourteen enemy planes with only eighty-eight rounds of ammunition.

One evening at the movies I bumped into the colonel's operations officer, Lieutenant Colonel Fitzgerald. We had been together at the Naval Academy Prep Class. After seventeen years we met again at a moving picture show on an island in the middle of the Pacific!

Another man on Eniwetok who was doing a job was Commander MacIsaac, commanding officer of the Carrier Aircraft Service Unit whose task was to keep the planes in operation. We had been to Pensacola together! "Before long," Commander MacIsaac remarked, "this two-by-four island will be as well known as Broadway and Forty-second Street!"

One day a report came in that a battleship had been seen in Truk Lagoon, southeast of Eten Island. That night, after covering my regular search sector through the Eastern Carolines, I headed for Truk to have a look.

The Japs evidently had an idea that we would be in to have a look. As we approached from the northeast, hugging the sea to avoid detection, a bright red flare blossomed in the darkness above the Fefan hills, coloring the sky for five long seconds before it sputtered out. My co-pilot, Lieutenant (jg) Norman Burton, said quietly, "That must be an air-raid signal."

I thought he was probably right. But it was a dark night, the moon veiled by a heavy haze, and obviously we had not yet been seen. On the other hand, we couldn't see much ourselves. The light was deceptive, and many of the smaller Truk islands looked like ships.

With a word to the boys to be on their toes, I flew low over the lagoon, searching in vain for the reported battleship, then turned south past the eastern end of the Dublon-Eten gap in quest of a substitute target. As we crossed the mouth of the gap we caught our breath. There in the funnel, about half a mile distant, were two

Jap men-of-war, the nearer one a cruiser!

There was not time to make a turn into the gap. Instead, I banked around the southern side of Eten Island and at two hundred feet above the water made a near-flipper turn to enter the funnel from the other direction. It was almost the last maneuver *Thunder Mug* ever made. Sharp turns to the right at high speed are always difficult, for the pilot sits at the left of the cockpit and customarily looks out the window on his own left. In trying to line up the target and bring us around at the same time, I misjudged our height above the water and let the plane start diving, neglecting my instruments in my eagerness to see the Jap and begin the bombing run.

Only Burton's presence of mind saved us. Watching the radio altimeter, he saw what had happened, grabbed at the yoke, and pulled us up just as we were about to plunge into the water at two hundred miles an hour.

By the time we had recovered from this close call, the cruiser was only half a mile distant, dead ahead, and our run had to be made almost straight down her longitudinal axis. In other words, we were in line with her keel, and if our bombs were even a little bit off on either side, she would probably suffer no great damage. There was not time for a more angular approach, however. At any moment the Japs might come to life down there and fill our path with a concentration of flak. I had no suicidal desire to face the full fury of a cruiser's AA batteries.

With full throttle we traveled the length of the ship, dropping three 1,000-pounders, then executed a radical turn to the left as the bombs exploded. All hit within ten to twenty feet of the target. One burst just off the ship's fantail. The other two lifted mountains of water close alongside. Our radio operator, stationed in the open bomb bay to keep the doors open, sang out a yell of elation.

Wondering why we had received no return fire from so potent a target, I came back across Eten airfield, strafing the runway and the installations there, and swung over for another look at our cruiser. She was still there. But now, listing badly, she was headed in almost the opposite direction. Our bombs had picked her up by the tail and turned her at least ninety degrees to the right. No wonder her guns were silent. Her gunners were probably piled up in a heap somewhere, trying to untangle themselves!

But if the cruiser's people were out of action, the AA gunners on Eten, Dublon and Fefan were not, and now the fire was the most intense we had ever faced. Most of it appeared to be medium stuff, 20-millimeter fire, and as we sped between the two islands it crossed just above us in a fiery arch, the tracers streaming over the plane from both directions. *A military wedding*, I thought suddenly. But it could have been a glittering, gaudy military funeral instead, if the boys of *Thunder Mug's* grand crew had not unsteadied the Japs' aim by giving them back as good as they sent. The muzzle blast of the fifties in our bow turret was so intense that I could not see through it without being blinded, and for a minute or two—a very bad minute or two—we were rumbling through that murderous cross-fire almost completely on instruments, jinking as violently as possible just above the water to keep the arch of tracers above us. Long after we were out of the gap, the Japs continued to fill that part of the sky with a greenish blue geyser of flame.

The AA fire followed us as we turned up the west shore of Dublon, strafing the seaplane base at Moen and a number of small naval auxiliaries at anchor there. It stopped, at least in our immediate vicinity, as I swung over the Moen air strip to give the boys a chance to do some damage there. Then we flew in peace and quiet up to North Pass, hunting a target for the single depth bomb we carried, and, failing to find one, we turned for home. One by one the crew came up into the cockpit to tell what they had seen, so that I could send the usual flash report, after which, if not on watch, they "flaked out" as the plane leisurely droned homeward through the night. Another mission was over. The Japs would be a long time putting their cruiser back into action. *Thunder Mug*, filled with flak holes, lumbered along as placidly as though she were returning from just another milk run—which, in fact, she was. Home was four to five hours away.

But I remember a comment made by one of the gunners, when he came into the cockpit. "Cap'n, it's funny about that flak from Dublon, Fefan, and Eten as we went through the gap," he said.

"That was just like the arch of swords at a military wedding." We had flown together so long that we were even beginning to think alike!

"How long do you think we were in Truk Lagoon altogether?" I asked him.

"Long enough!"

"Just eighteen minutes," I said, and Burton nodded his corroboration. In eighteen minutes we had undoubtedly cost the Japs many thousands of dollars, not to mention the time and labor needed to repair the damage done by our lone Liberator. But figures are misleading. In order to spend eighteen minutes inside the enemy's fortress, we had been dependent on *Thunder Mug's* four engines for more than fifteen hours of flight, and behind the flight lay hours of the most patient planning. A lot of hard, sticky work had gone into the battering of that Jap warship. Eleven men made up *Thunder Mug's* crew, but many times eleven men had contributed to the success of the mission. The headline writers ought to remember that.

[*The probable destruction of this enemy cruiser in the Truk inner fortress won for Commander Miller another Air Medal, the citation reading as follows: "Upon sighting an enemy light cruiser and another vessel lying at anchor in the lagoon between Eten and Dublon Islands, Commander Miller immediately maneuvered his plane into an advantageous striking position and heroically flew down the length of the cruiser at masthead level, dropping three bombs which hit close aboard and leaving the hostile ship heavily damaged and listing. The attack completed, he made a daring strafing run on the airfield installations at Eten and subsequently proceeded to Moen Island, accurately strafing enemy vessels, ground installations and a seaplane base enroute despite intense antiaircraft fire from Eten and Dublon. Commander Miller's splendid initiative, superb flying skill and tenacious devotion to duty under extremely difficult conditions were in keeping with the highest traditions of the U. S. Naval Service."—H.B.C.*]

PRAYERS AT PULUWAT

ON the first day of May a slender, red-haired young lieutenant walked into my tent and announced that he had been sent to Eniwetok to replace Ed Lloyd as our air combat intelligence officer. Many a time since, we have laughed over my reaction.

"Your mouth dropped open," Ted insists, "and you looked at me as though I were something in diapers, representing the New Year!"

I probably did. Ed Lloyd had been our intelligence officer since the start of the tour, and we owed much of our success to his careful, brilliant work. The thought of having this wise old veteran replaced by an apparently green youngster was naturally a shock. An incompetent ACIO can wreck a squadron as surely as a good one can keep it on its toes.

I needn't have worried. Lieutenant Ted Steele was a Dartmouth man and a Rhodes scholar. He had been with old VP-82 when that squadron, based in Newfoundland and Greenland, operated on the Eastern Sea Frontier—back in the days when our Navy was so short of patrol planes that the British sent over one of their squadrons, RAF-57, to lend us a helping hand. From there, old Ted had gone to torpedo planes as a communications specialist, then to the staff of ComAirPac. When Ed Lloyd departed four days after Ted's arrival, our new ACIO was already in the groove.

Ed, a much loved guy, got off before dawn while the rest of us were still asleep. This is the note he left:

"Bus and George and Jan and John and Tommie and all pilots and co-pilots as well as navigators and goddam ground officers: Good-by, you guys. I'm glad the plane leaves early so I can sneak out alone like this. Give 'em hell and God guide you softly. Ed."

Ed really knew how to spell *safely*. But a girl who had been reading about us in the papers back home, and had written us a letter telling us how very much she loved us all, apparently didn't. Her "God guide you softely" had become a squadron by-word.

Except for a daily two-plane reconnaissance of Wake, and the usual patrols, these early May days were uneventful. The boys spent their spare time playing poker or trying to track down a battered copy of something called *The Chinese Room* which had mysteriously appeared on the island. We griped at the food, went to the movies occasionally, flew far beyond our assigned search sectors to be sure the Japs were not up to something, and began to wonder when the advance on the enemy's homeland would be continued. The Japs at the moment appeared to be husbanding their shipping carefully. Our targets were few.

One day, however, Lieutenant Keeling stirred up a flurry of excitement by tangling with a Jap Mavis (a four-engined seaplane) and shooting it down in flames after a nineteen-minute chase in and out of the clouds. It was the first enemy plane we had destroyed in the air. Donald Maloney, Keeling's photographer, brought back a set of some of the finest action photographs ever taken. And on a pre-dawn solo mission to Truk, *Thunder Mug* was instrumental in saving the lives of three Army airmen whom we found floating on a rubber raft west of Truk Lagoon.

In an effort to do something about the shortage of shipping, I made several trips to Truk about this time—even there the shipping was scarce—and also to near-by Puluwat, where I had discovered a new and important enemy airfield under construction. It was at Puluwat, on May 16, that I almost came to grief.

On a mission over Truk two days before, I had seen some Jap ships in the anchorage and was reluctant to let them depart before working them over. My crew, however, was tired. For weeks I had been promising them a deserved rest. This time, keeping the promise, I left them at home and took Lieutenant Bill Bridgeman's plane, *Climbaboard*, with Bill as co-pilot.

They were not expecting us at Truk that night. Of ten to twelve cargo ships at anchor in the old hunting ground, we attacked the

largest, tossed it around with three bomb explosions, and were out
of range before the surprised Japs could use their guns with any
effect. Our own port waist gunner, however, shouted that he had
seen a bigger target against the moonlight over near Dublon, so I
flew back through the Fefan-Dublon corridor. By then the Japs
were awake and peeved. From half the islands in Truk they were
hunting us with intense and uncomfortably accurate fire. But the
target, a 10,000-ton tanker, was too rich a prize to be passed up.

It was the kind of run a bomber pilot dreams about. At 200 miles
an hour, 200 feet above the moonlit sea, I pickled off three bombs
as the ship appeared broadside in the sight. The first struck the
water fifteen feet from the tanker's hull. The second exploded
squarely on her deck; the third was over. An instant later the whole
sky behind us was aflame and the tanker, evidently laden with oil,
exploded on the sea like an active volcano. Nothing else makes quite
the sound and fury of an oil explosion.

By now, though, the Japs had us in their searchlights, and the
defensive fire from the islands was nearly as brilliant as the blaze
from the burning tanker. We escaped through Piaanu Pass, leaving
them shooting at shadows, and set our course westward to Puluwat.

I have purposely avoided much mention of Puluwat until now,
in order to keep this account from running off in all directions
across the Pacific. But we are about finished with Truk, and Puluwat may be given its proper importance. The atoll lies some 120
miles west of Truk, and was about at the extreme end of our search
from Eniwetok. There is not much to the place. Of the two main
land masses, Alet Island is but a mile and a half long and Puluwat
Island, across the lagoon, is not much larger. But on Alet the Japs
were clearing an air strip which could be used as a satellite to the
fields at Truk, and on Puluwat they had set up efficient weather
and radio stations. And, of course, there were Japs on both islands
to attend to these things. (See map, p. 157.)

It was coming on daylight when *Climbaboard* arrived over the
Alet air strip after the run from Truk. A group of Japs had just
emerged from their tin-roofed shacks and were walking down the
strip on their way to work. They were considerably less leisurely

in their strolling when the big Liberator, appearing suddenly over the treetops, bellowed down on them and sprayed the strip with bombs and bullets. A truck, zigzagging like a frightened chicken, sped straight into a line of tracers from one of our turrets, and burst into flames.

We flew on, dropping 500-pounders singly on individual targets, strafing the radio and weather stations, the barracks, and a pre-war

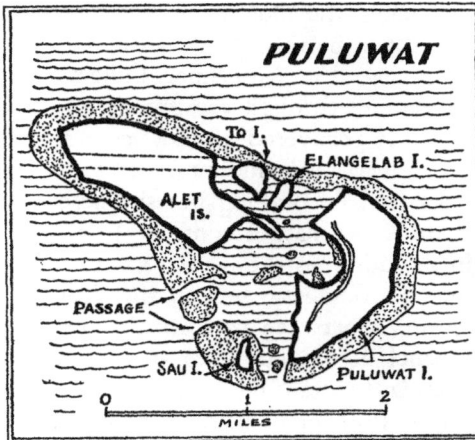

lighthouse, 135 feet high, in the top of which the Japs had installed a potent 7.7-millimeter gun. In the dawn dimness their gunners had difficulty finding us, and even more trouble following us as we made our runs. But that lighthouse gun was annoyingly accurate, and after damaging the radio station with two direct bomb hits, we went after it. For a few moments the Pacific war was reduced, insofar as we in *Climbaboard* were concerned, to a private duel between a Liberator and a lighthouse.

The Liberator won. On our last pass—so low over the tower that the belly gunner wailed up through the interphone, "For Pete's sake, Cap'n, you're scraping my duff on this thing!"—the plane's gunners put the tower out of commission. Then we jinked away, avoiding the ground fire, to make another run on the radio station.

And right there a Jap three-inch shell burst above *Climbaboard's* cockpit.

The plane bounced like a big rubber ball. Old Schaffer, standing between the pilot and co-pilot, was knocked off his feet by the concussion and tumbled backward out of the cockpit to the catwalk below, where, but for our habit of keeping a man stationed at the bomb bay doors, he would have gone right on out. But Bridgeman and I were not aware of Shafe's close call until later. The explosion of the shell had kicked the blocks out from under the top turret guns and depressed the barrels. Now, with the gunner dazed and unable to cease firing, the .50-caliber bullets from those lethal muzzles came streaming through the greenhouse into the cockpit.

Luckily the gunner recovered after a second or two. But twin fifties can spit out a terrifying amount of steel in a few seconds, and Bridgeman and I could do nothing but sit there. Bill sat with one hand frozen on the supercharger controls. I held the pickle in one fist, the yoke in the other, as the plane recovered from the shaking up and continued its bombing run on the Puluwat radio station. The bullets, screaming between us from the turret above, smashed the instrument panel before our eyes and filled the cockpit with flying bits of glass and metal. The muzzle blast singed the hair on our heads and arms. Both of us bled from scores of small wounds.

The plane, meanwhile, had continued its run, and from habit I pressed the pickle as the radio station rode into the sights. It was just "one of those things." Four 100-pound bombs fell squarely on the building. Then, with the pilot and co-pilot dazedly fumbling at the cockpit controls to find out how many of them were still functional, we turned for home.

That was a grim ride. With old Shafe helping us, we spread sulfa powder on our wounds and swallowed tablets to dull the pain, not daring to pull the fragments of glass and steel from our flesh for fear the bleeding would increase. One of us, at least, had to stay conscious long enough to put the plane and its crew safely on the ground at Eniwetok, 850 upwind miles away.

We were seven hours getting there. I want very much not to be dramatic about this, but seven hours can be a very long time. It will

take you from Boston to New York by car, or see you through
nearly a full day at the office. But for seven hours, in a sulfa stupor,
we sat at the controls of a badly shot-up Liberator, bucking a ten
to twenty knot head-wind over the empty Pacific, while nine good
men prayed quietly that one of us would hang on long enough to
take us in.

There were times during that long grind when we thought we
were not going to make it—when one or the other of us began to
imagine that the plane was doing queer things. But we landed
safely at base seven hours after leaving Puluwat.

[*A fellow officer noted something else, too. "Recovering from
the initial shock and daze, and discovering that the plane would
still fly despite the cockpit carnage and instrument wreckage, and
that he still had control of it despite his many injuries, Miller
RADIOED BACK THE WEATHER REPORT and returned to
his search sector, landing safely at base seven hours later." Capitals
mine.—H.B.C.*]

[*In the name of the President of the United States, the Commander takes pleasure in presenting the NAVY CROSS to*
COMMANDER NORMAN MICKEY MILLER, UNITED
    STATES NAVY *for service as set forth in the following*
    CITATION:
    "For extraordinary heroism in action against enemy Japanese
    forces in the Caroline Islands, on 16 May 1944. As pilot of a
    PB4Y Liberator bomber, he flew a long range search mission
    through the northern Carolines, then proceeded to Truk to
    investigate shipping anchored in the lagoon. He immediately
    attacked a five thousand ton cargo vessel at masthead height.
    Flying from stem to bow he dropped three bombs close aboard,
    heavily damaging the enemy ship. He then made a similar at-
    tack, under increasingly intense enemy fire, on a ten-thousand-
    ton tanker, which resulted in a great explosion which com-
    pletely destroyed the large vessel. He then proceeded to Pulu-
    wat, still at minimum altitude, where he surprised a party of*

*Japanese on the airfield and by strafing killed more than thirty
of them and demolished their truck. By a direct bomb hit he
completely destroyed their revetment. He attacked a radio
station, obtaining two direct hits. While he was making a second successful attack on the radio station, during which he
obtained four direct hits, a Japanese antiaircraft shell burst just
above the plane, nearly demolishing the cockpit and wounding
both Commander Miller and his co-pilot. Despite shock and
profuse bleeding, and with his plane severely damaged, he flew
the plane safely back to base, over eight hundred miles distant.
His cool courage, his consummate skill as an airman, his outstanding devotion to duty and aggressive fighting spirit were
an inspiration to the squadron under his command and in keeping with the highest traditions of the naval service."*

J. H. Hoover
Vice Admiral, U. S. Navy]

Temporary Citation.

21   *FIRST TO SAIPAN*

On Apamama, early in the tour, we had set up a board for the
recommendation of awards, with Lieutenant Commander George
Hicks, the squadron executive officer, as chairman. No one, of
course, was over-eager to serve on such a board. The appointees
had to be brave and honest souls, free from bias, impervious to
criticism, with more than a passing knowledge of psychology and
philosophy and the ability to bear up under an avalanche of paper
work.

They performed their duties well. Moreover, they managed to
keep their sense of humor. Toward the end of May, Yeoman Chester Jones solemnly presented me a paper for my signature and said,
"Commander, the forgotten men have risen in revolt. Peruse this,
if you please."

This is what he had written:

Recommendation for awards to Personnel for conduct during operations with the Board of Awards, U, S. Pacific Fleet.

1. On 2 February 1944, ComBomRon appointed a Board for the recommendation of awards. In the order establishing this Board its duties were set forth as follows:

(a) Decide on the merits and appropriate award for any acts of heroism and gallantry in aerial flight, or of heroism and gallantry not involving combat, or awards for wounds in action against the enemy.

(b) Prepare all paper work involved in recommending these awards.

(c) Take cognizance and appropriate action in all matters pertaining to awards within this squadron.

2. Immediately upon receipt of the order stated in (1) above, the members of this Board, with utter disregard for the inevitable criticism to be leveled at them, and knowing full well that they would be engulfed immediately in a murderous cross-fire of superlatives, did calmly, courageously, skillfully, adeptly, adroitly, persistently, aggressively, daringly, intensively, accurately, conspicuously, bravely, lethally, perfectly, brilliantly, audaciously, precisely, substantially, co-operatively, destructively, spectacularly, devastatingly, dramatically, effectively, severely, heavily, directly, surprisingly, superbly, excellently, unusually, doggedly, completely, intelligently, and successfully proceed to evaluate with great determination, initiative, and dexterity the exploits of their fellow officers and men of Bombing Squadron ONE HUNDRED NINE.

3. Working against time (GCT) and with no thought of personal fatigue or hope of credit or award, and indeed, with the prospect of receiving nothing but abuse and recrimination, the members of this Board during the period from 2 February 1944 to 23 May 1944 inclusive, brought their mission to a successful, favorable, advantageous and prosperous conclusion.

4. As a result of the actions stated in (2) and (3) above, it is recommended that the Legion of Merit be awarded the following:

(a) Lieut. Comdr. George L. Hicks, USNR, for exceptionally meritorious conduct in the performance of outstanding service while participating as chairman of the Board of Awards, Bombing Squadron ONE HUNDRED NINE. From 2 February 1944 to 23 May 1944 inclusive, Lieut. Comdr. Hicks served indefatigably, assiduously, indomitably, industriously, perseveringly, persistently, tirelessly, and with unfailing, unfaltering, unflagging and unwearied zeal to successfully recommend the award of 277 decorations to the officers and men of Bombing Squadron ONE HUNDRED NINE. His disregard of exhaustion and his devotion to duty were in keeping with the highest traditions of the Naval Service.

(b) Lieut. Comdr. John F. Bundy, USN.

(c) Lieut. Comdr. William Janeshek, USN.
    The citations for 4(b) and 4(c) are the same as for 4(a) above, except for the substitution of the word "member" for the word "chairman."

(d) Jones, Chester Smith, USNR, for exceptionally meritorious conduct in the performance of outstanding service while serving as yeoman to the Board of Awards, Bombing Squadron ONE HUNDRED NINE. From 2 February 1944 to 23 May 1944, inclusive, Jones, in spite of an almost overwhelming amount of paper work, and in the face of execrable penmanship, did faithfully and with trustworthy zeal, translate, correct, polish and type recommendations for 277 awards to members of Bombing Squadron ONE HUNDRED NINE. His disregard for operational fatigue and his devotion to duty are in keeping with the highest traditions of the Naval Service.

(e) Sladich, John S.
    The citation for 4(e) is the same as for 4(d) above.

One day Lieutenant Harold Belew, who in addition to being a patrol plane commander was also our material officer, delivered a speech and presented me a medal of his own making. For a long time I had been misplacing flashlights, which were extremely valuable items during night flights and blackouts. Flashlights on Eniwetok were scarce, and Belew was becoming desperate. So, when at last I managed to go a whole week without having to ask him for a new one, he obtained a safety pin, some bits of red ribbon, part of a Kodak film carton, and a metal disc stamped "Consolidated Aircraft," and fashioned a medal in honor of the occasion. I never again lost a flashlight. Well, I almost never did.

But these were busy days. In June, our Army, Navy and Marine commanders were completing their preparations for the invasion of the Marianas, next stepping stone on the road to Japan. The Marianas had been scouted; they had been hit by carrier-based aircraft. Lieutenant General Alexander A. Vandegrift, Marine Commandant, was on record with the statement that we stood on the threshold of bigger and more difficult landings than we had ever yet attempted.

As the invasion forces gathered, our already heavy schedule of patrols was expanded, and we paid particular attention to the increased amount of enemy shipping at Truk. Scouting the atoll on the night of June 2, I discovered so many choice targets that it was hard to choose among them. We blew a 7,000-ton cargo ship almost out of the water with a direct hit, worked the airfields and seaplane base over, found a dozen smaller ships at the northwest end of the Dublon anchorage and damaged five of them, traded machine-gun fire with a destroyer, and wrecked a plane at Eten.* Yet with all these ships in Truk, we had been singularly unsuccessful in locating them on the high seas. Obviously they were moving into Truk at night, along some shipping lane we had not yet discovered. If so, they were a menace, for our fleet had to pass west of Truk to reach the Marianas.

The following night Lieutenant Commander Hicks did locate the enemy's new shipping route, far west of our usual patrol limits, and sighted a convoy of three merchant ships, seven armed naval auxiliaries and a destroyer. Despite extremely poor visibility he gave them a thorough drubbing, heavily damaging two of the cargo vessels, three of the escorts and the destroyer. But the plane, caught in cross-fire over the center of the convoy, was badly holed. And the plane captain, Hale Fisher, working the starboard waist gun, was seriously wounded. The 12.7-millimeter shell that wounded him entered the fuselage under the pilot's seat and traveled the length of the plane through the bomb-bay tank and two metal bulkheads before striking Fisher in the stomach.

* "*The audacity of this foray and the damage inflicted,*" *wrote Captain Taff,* "*are outstanding.*"—H.B.C.

The navigator, Ensign Dicky Wieland, ran aft when Fisher was hurt, gave the wounded man morphine, bandaged him, and remained at his side all the way home. In the words of the flight surgeon who took over at Eniwetok, when they landed at 1 A.M., "no medical man could have been more faithful or capable." But when Admiral Hoover and General Hale presented their first awards to some of the boys of VB-109 and VB-108 the following day, with the band playing and the sun shining warmly on the crowded air strip, Hale Fisher was not there to receive his air medal.

The following night, continuing this intense search for enemy shipping, Lieutenant Wheaton found a Jap convoy west of Puluwat, out of which he sank a gunboat and crippled two escorts. After which Lieutenant Mellard picked up a convoy and got himself a fat cargo ship and a possible destroyer. The Japs were really on the move.

But American forces were on the move, too, and as the date of the invasion of Saipan approached, VB-109 worked overtime. Information concerning the size and disposition of our forces had to be kept from the enemy at all costs, lest the Japs find time to prepare a counter blow. All over the Central Pacific, groups of our ships were on the move—tankers, LSTs, carrier forces, battleships, cruisers, destroyers, and a bewildering array of invasion craft. Day by day we patrolled ahead of them, relieving one another on station as our fuel ran low. Our job was to be sure that no enemy shipping lay in the path of the advance, and to knock down any Jap search planes that might give the show away.

On one occasion during this patrol, Lieutenant Commander Janeshek encountered a Jap torpedo bomber within sight of Truk and shot it down in an encounter lasting less than five minutes. As the Kate plunged toward the sea with Janeshek's gunners still working it over, the Jap rear gunner frantically threw back his canopy and stood up to bail out. "He suddenly sat down again," reported Janeshek gravely. Five hundred feet above the water the Jap jettisoned two depth bombs, which exploded beneath the plane as it struck the sea. "Napping Nip Now Nil," Janeshek reported.

Not to be outdone, Hal Belew jumped one of the Jap's newest

twin-engined fighters, an Irving, and sent it home smoking. Then it was George Mellard's turn.

Mellard, patrolling just ahead of the fleet with Lieutenant Hop of VB-108, was flying at 9,000 feet above a solid overcast when Hop, below him, flushed a two-engined Jap Betty. Hop jettisoned his bombs and went after the enemy plane at once, but the Jap, too smart to invite a shooting match, turned tail and ran up into the overcast. By radio Hop warned Mellard above, and Mellard was waiting. Using just 100 rounds of ammunition and a minute and a half of time, Mellard's gunners sent the Betty down again, this time for keeps. Then, as the Liberator resumed its patrol, fighters from a carrier force only fourteen minutes distant came up to give the boys a nod of thanks.

(I should include here a little incident that occurred when Lieutenant Hop encountered a Jap Betty during our stay at Eniwetok. This is what we heard as we listened to him on the base radio:

Pilot (being very formal): "Have sighted Betty. Am trailing."

A few minutes later (still very formal): "Have contacted Betty. Am engaging."

A moment later (a gleeful shout): "I dood it!"

One more little story: Lieutenant Oliver Glenn's Liberator, *Urge Me*, carried an exceedingly luscious nude on her fuselage. During our pre-invasion patrol, one of the carrier pilots discovered this and flew alongside for a look. He obviously liked what he saw, for over the radio he invited his buddies to the party, and in perfect formation, one by one, they came alongside for inspection. Then, with wide grins of appreciation, they departed.)

Photographs are of vital importance in any venture such as the invasion of the Marianas. Without them, target maps cannot be corrected to reveal the latest enemy gun positions, radio stations, troop dispositions, and a score of other items which must be known in advance. Weeks before the fleet was assembled for the attack on the Marianas, photo planes of VD-3, based with us at Eniwetok, had begun their high-altitude visits to the islands. Sometimes we accompanied them to drop a few token bombs.

But an enemy cannot be expected to let his defenses remain stagnant, and photographs quickly become outdated. Even while the fleet was steaming to the attack, new pictures had to be delivered to Admiral Raymond A. Spruance and Vice Admiral Richmond Kelly Turner, in whose hands the fate of the invasion rested.

This was a job for VB-109. As fast as the photographs were handed to us at Eniwetok, I loaded them aboard *Thunder Mug* in sealed pouches and set out over some nine hundred miles of ocean to deliver them.

For hours, as we flew in solitude over an empty sea, the war would seem as distant as Denver. Then suddenly all the ships in the world would appear beneath us, their white wakes weaving an intricate tapestry as they plodded westward. To one who remembered flying the old XPBS to the shambles of the Southwest Pacific just after Pearl Harbor, it was a thrilling spectacle.

The larger ships were unable to maneuver from formation to pick up our pouches, and so I usually singled out some small escort vessel, made a pass or two across her bow, and then, flying low over the water ahead of her, called out to the boys to "let 'em go." As we dropped the pouches, the attached float-lights spouted smoke and flame to mark their position. Then, after waiting around until the containers were fished from the sea, we waved farewell and continued our usual search assignment.

The naval bombardment of Saipan began June 10 and lasted four days, while carrier-based planes gave the island a pre-invasion drubbing. On the 14th, American troops hit the beaches.

On the 17th, in answer to the admiral's request for more photographs, Lieutenant Commander Janeshek flew low over the Saipan defenders and obtained excellent low-level shots for use by our assault troops. Two days later Lieutenant Jobe covered Tinian Island as well.

On the 29th, with the battle for Saipan still in progress, Lieutenant Tom Wheaton took his cameras on a similar mission and, after snapping his pictures, put his plane down on newly captured Aslito airfield. It was the first heavy bomber to land there. No doubt it occasioned some lifting of eyebrows when Admiral Hoover, General Hale and other officers of high rank arrived an hour later in a Liberator transport, expecting to find nothing but fighter planes on the Aslito air strip.

22   *HIGH JINKS AT IWO JIMA*

OUR troops had occupied most of the southern part of Saipan by the first week of July. But they by no means had all of that mountainous little rampart in the Marianas, and there remained a number of excellent targets to sharpen our picture-taking sorties.

The most important of these was the airfield at Marpi Point, where the Japs had installed powerful AA guns in sunken pits. On July 1, and again on July 5, I laid strings of bombs along this field while *Thunder Mug's* crew concentrated on the pits. The guns were no longer blink-blink-blinking at us when we left, nor were they in condition to blink at the Marines who came up through the mountains a few days later.

Another prime target was the Ushi Point airfield on the northern tip of near-by Tinian, about two miles south of Saipan across the channel. The enemy there had erected concrete fortifications to repel invaders, and our warships were working them over in an effort to lighten the chore for the troops who would soon be hitting the beaches. *Thunder Mug* finished the job by rumbling in over the hangar roofs and the air strip and blowing the reinforced concrete to rubble with a string of bombs.

Saipan, however, was a long reach from our base at Eniwetok, 100 miles beyond our patrol, and our effectiveness on these volunteer "extra-curricular" missions was reduced by the limit placed on our bomb loads. It was further reduced by the presence of American troops on the island, some of whom were not always where we were told they would be, and by the many American ships which lay at anchor in near-by waters. Before dropping any bombs at all we had to fly low over the fighting, usually through rain and smothering clouds, studying the bombardments, the tank actions, the brilliant spurts of fire from the flame-throwers, to determine the exact location of the front lines. Only then could we safely aid the ground troops by attacking the enemy from the air.

For that matter, at this point in the campaign there was not much we could do anywhere to contribute to the winning of the war. Truk, though still a fertile hunting ground on occasions, was being by-passed and would be of little use to the Japs after we had established ourselves in the Marianas. Ponape and Kusaie were all but impotent, Puluwat had nothing more to offer, and in our routine search areas we encountered no enemy at all. So, with the sincere conviction that we were wasting our time at Eniwetok, I appealed to Admiral Hoover's operations officer, Commander "Bucky" Lee, to get us out to Saipan. From Aslito Field, which was now safely in American hands and had been renamed Isley Field, we could do business northward in the Kazans and Bonins.

On July 11 my request was granted. There was not room for all of us at Saipan, of course, but from Captain Taff came permission to take two planes. Late the following afternoon Lieutenant Joe Jobe and I, with our crews, landed at Isley Field.

It was a dismal place. The pre-invasion plastering had been severe, and continued heavy rains had made difficult the task of reconstruction. We, of course, were orphans. This was a fighter field, with no accommodations whatever for heavy bombers, and it was an Army field, to boot.

But the Army's operations officer, Captain Fahle, though already hard pressed to accommodate three squadrons of P-47 fighters, gave us a warm welcome and found us a place to park our two Liberators. In addition he contrived to locate for us a barrel of water. God knows there was water enough on Saipan—it was ankle deep everywhere—but you couldn't drink it. Every blessed thing on that muddy, mountainous, fetid little island was a source of potential sickness. The mud never dried, the rain seldom stopped, the flies and mosquitoes never slept. Our troops, before landing, had been warned not to eat, drink or do anything that was not strictly G.I.

Our arrival at Saipan did not prove anything, of course. I still had to obtain my direct and immediate operational orders, and then would have to prove that we could get a four-engined bomber off the primitive Isley Field fighter strip with bombs enough to do the enemy some damage. So while the boys bedded down for the night as best they could in the mud under the wings of the planes, I went looking for Captain Taff, our operational boss.

That was one pilgrimage I won't quickly forget. The captain, I was told, was aboard a ship "somewhere off shore." I rode in a jeep, through dripping darkness, along what had once been a road of some sort, but the jeep kept wandering off the road into the bordering cane fields. These cane fields, a source of raw material for the island's sugar mills, were breeding grounds for hordes of man-eating mosquitoes which rose in angry clouds when disturbed. And there were Jap snipers in the fields, too—but they must have been too busy battling the mosquitoes that night to bother with me.

Eventually I located a boat at the beach, and my "journey through the dark" became a voyage, but the change was not an improvement. I kept finding ships but not the one I was looking for. Three full hours had passed before I discovered her at last.

Captain Taff knew what I wanted; he had approved our coming to Saipan in the first place. In fact, I had among my effects a letter from him, dated June 8, in which he had written: "It is noted that your frequent forays against strong enemy bases and convoys have met with singular success, and that your methods have in some cases been passed on to other members of your squadron and have thus enabled them to achieve similar success. It is therefore requested that you prepare and submit, in as much detail as necessary, a treatise outlining these methods . . . in the hope that with such information, future squadrons may achieve comparable success." My hope now was that the captain would say something as nice as this to Admiral Spruance and persuade the admiral to let us go soon on a prowl among the Jap islands to the north.

"Better wait till morning," Captain Taff advised. "Then I'll go with you to the admiral and we'll talk to him."

The admiral, when I met him, was an agreeably easy man to talk to. His clear blue eyes impressed me quickly. So did his adherence to essentials and the precision of his line of thought. He had heard of us, and agreed that if we could get to the Bonins and the Kazans before the Japs anticipated a strike by land-based aircraft, we might do some worthwhile damage. Iwo Jima, in the Kazans, would be of tremendous importance to the Japs now that they were losing Saipan. Almost certainly they would begin at once to reinforce it as a citadel to protect their homeland. We should find some valuable installations there, and if we struck quickly we might pick up some enemy shipping, in which the admiral was particularly interested.

"Go ahead," he said. "Look in on Iwo Jima and the Bonins and see what they've got there. Go after their destroyers especially. Good luck and good hunting."

Next morning Joe Jobe and I, with our co-pilots and navigators, studied maps and photographs of Iwo Jima until we knew every hill and harbor by heart. In the pictures, the island looked like an elongated turtle with its head pointing roughly to the south, its tail aimed in the general direction of Japan. This Jap turtle, its flippers enfolding the island's two anchorages, its mountainous back supporting two good airfields, lay 625 nautical miles north of Saipan and only 670 nautical miles south of Tokyo. A few miles north of it lay Kita Iwo Jima; a few miles south lay Minami Iwo Jima—both too small and too precipitous to be of much importance except as guardian outposts for Iwo Jima itself. The three islands, taken together, comprised the Kazan Group of the Nanpo Shoto.

We decided to hit Iwo Jima at dusk, and so at two o'clock in the afternoon the planes were taxied to the extreme end of the Isley Field fighter strip and readied for the take-off. Each crew had poured two thousand gallons of gas into the tanks, enough to get us there and back, and hung two thousand pounds of bombs on the racks. We were going to have to take off in the rain, with no wind, from a runway just 3,800 feet long. Skeptical Army pilots lined up along the strip and ground crews stopped work to watch in silence as battered old *Thunder Mug* turned into position. They didn't know our PB4Ys as well as we did.

The crew was ready. Old Shafe stood between me and my co-pilot, hairy-chested Jim Park—"Alley Oop" we called him. "Let's

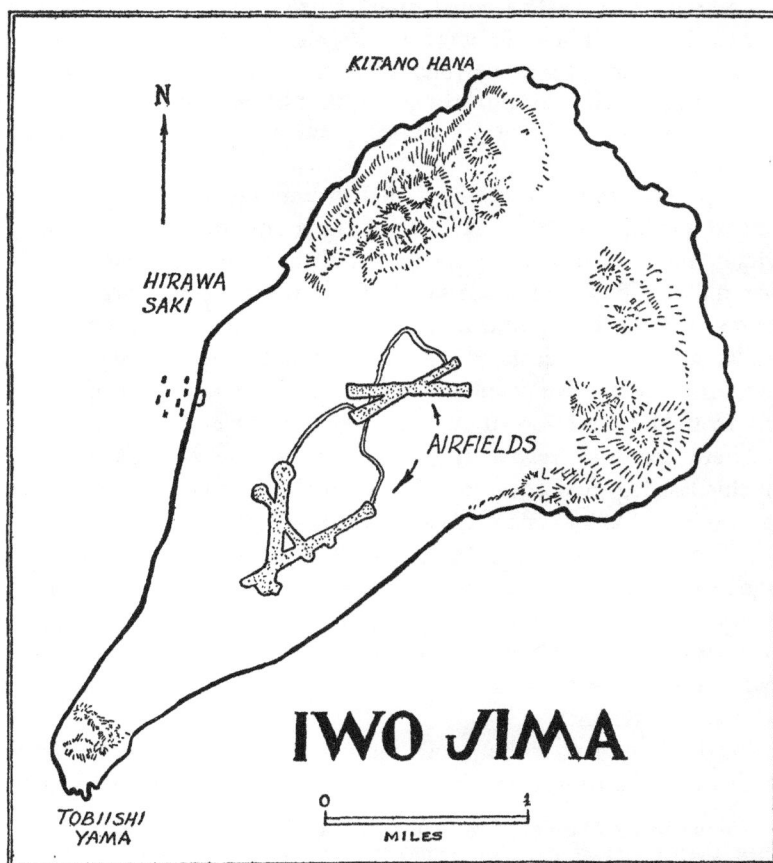

KITANO HANA

N

HIRAWA
SAKI

AIRFIELDS

IWO JIMA

0   1

MILES

TOBIISHI
YAMA

make it good, Skipper," Shafe begged. "Just to give those Army Joes the laugh." I turned the engines up full, seven inches beyond maximum manifold pressure, and released the brakes. The plane went lunging down the lumpy field like a hound dog off a leash.

With her flaps up, for speed without drag, she roared along until we were doing an uncomfortable 100 miles an hour. Then I lowered the flaps for lift, and *Thunder Mug* tipped her nose up. For a moment her wheels seemed glued to the ground as she strained to free herself. Then we were in the air. Later we learned that a cheer went up behind us, but with the four engines thundering like freight trains in our ears, we didn't hear it. We ourselves cheered, however, when Joe Jobe's plane swept up from the field in our wake.

Keeping close together for added protection, the two Liberators ranged far up to the north, shunning a number of small islands which might have been outposts for the enemy's air-raid warning system. We flew on past Iwo Jima as planned, and arriving too soon at our turning point, had to loaf around over the Pacific for a while, waiting for dusk. When the sky had darkened and the sea had turned gray, we went on in, approaching not from the south but from the direction of the Japs' own homeland.

There was little talk as the two planes skimmed over the water on this last leg of the journey. This was big business; it might even be too big for our pair of unescorted Liberators, and the boys knew that our chances of getting back to Saipan with a whole skin —not to mention a respectable score—would depend largely on our ability to catch the enemy napping. There would be no lingering over this target, no doubling back to deal the foe a second dose. We had been warned soberly that Iwo Jima was bristling with powerful defenses.*

So, then, we had to be good—we could make no mistakes—if we were to prove the point I had so strenuously argued in my cam-

---

* United States Marines learned just how strongly defended Iwo Jima was when they assaulted the island on February 19, 1945. Every yard of the bloody little island was fanatically defended by an estimated 20,000 enemy troops. Marine casualties in the first three days alone were more than 600 killed, 4,000 wounded—heavier than at Tarawa—and the Fleet Marine Force Commander, Lieutenant General Holland M. Smith, called the Iwo campaign the toughest Marine fight in 168 years. "Iwo Jima," reported *Time* Correspondent Robert Sherrod, with the assault forces, "is probably the world's most heavily defended island."

paign to get just two of our planes to Saipan. Admiral Spruance and Captain Taff would be keeping an eye on us. We still had to convince them that by using a short, undeveloped advance field and flying far out beyond the radius of single- and twin-engined planes, we could do some important damage to an enemy not yet prepared for attack by heavy aircraft. Of particular importance would be our score against the enemy's lifeline of shipping.

Iwo Jima rises from south to north in a gentle slope which terminates abruptly in a bluff at the northern tip. The bluff towered before us, black against the sky, as we went in just over the sea. We hoped it would hide us. But we could not know whether it would or not until the two planes climbed from the sea, cleared the bluff, and swung down over the air strips beyond.

We were about eight hundred feet apart, Jobe on my port beam, as we cleared the barrier. *Thunder Mug's* gunners anxiously awaited the signal to open fire. "Okay, give 'em hell," I said—and twenty machine guns, ten on each plane, opened up with a spectacular, deafening racket, pinning the Japs to the ground ahead of us as we attacked.

The Japs were surprised, no doubt of that, but they were not panicked. The breed on Iwo Jima didn't scare easily. Our own yellow-red tracers had scarcely begun to ricochet from the ground when the greenish-blue ground fire began to flicker up at us. Only our speed and the surprise of the attack saved us. At two hundred miles an hour, barely 150 feet above the defenders' heads, we were fleeting targets despite our size. These Japs had never before had occasion to fire at anything so low. Their only previous targets had been carrier planes, at higher altitude. Nor had they ever been hit before by multi-engined aircraft. This was the first time Liberators had struck so close to Tokyo.

A few bullets nicked *Thunder Mug's* metal skin as we bore down on the first of the two airfields. Most of the flak, however, exploded in our slipstream, and the fragments that nicked us didn't hurt us.

No one except the Japs will ever know the full amount of damage done to their fields and parked planes that evening. At least thirty, probably forty planes were parked in the revetments and on the taxiways. Some exploded as our streams of machine-gun fire walked up and over them; some burst into flames; others spat out gusts of smoke. Over the interphone came brief reports from our gunners, but they were as busy as drunks in a shooting gallery and had little time or inclination to assess the damage already done. And by now we were really getting a reception. The Jap ground guns all over the island were putting up a barrage so intense that our encounters with enemy ack-ack in the Marshalls and the Carolines seemed tame by comparison. Even the ships off-shore were doing their Sunday best to knock us down. The sky ahead, behind and around us was aflame with gaudy Jap-made lightning.

I was saving the bombs for shipping, and as yet had released none of them. Our machine guns had been able to take care of the jammed planes on the airfields. But now, turning sharply to starboard, I headed for the anchorage on the west side of the island. With fires spreading on the ground behind us, I singled out a destroyer a mile and a half off shore and began a run on her. She was not the only target available but she was the best one, for the Japs were short of destroyers and desperately needed them for escorting their convoys.

This one had been firing at us even before we began our run, and, of course, she had weapons more powerful than anything with which we could answer her. Too smart to waste ammunition, the boys concentrated on the coastal vessels and cargo ships that came in range as we went in. To port, a cargo ship blazed on the water. Close aboard on the starboard side, a wooden coastal spouted flames. It was a good show. But Jim Park and I, in the cockpit, missed most of it. Enemy fire was bursting all around and over us by now, and we could see what most of the crew could not—the destroyer was sweeping the sky in front of us with twin streams of tracers from 20-millimeter deck guns. These twin rivers of fire rose in a rainbow arc between her stacks, so bright they fascinated me, and we were running down them like a train on tracks.

It was the longest, grimmest approach I had ever made, and I almost turned away. We were being hit hard; the Jap had a no-deflection shot and couldn't miss. And for some reason our bow gunner, who should have been doing the talking for our side, had failed to answer back. I didn't know until later that his guns had jammed and he couldn't!

I thought we would never make it, but old *Thunder Mug* drilled on through all the Jap could send, and the ship at last was in the sights. When I pickled off *those* four bombs, I really meant it!

No one had to guess at the results. *Thunder Mug's* gunners saw what happened, and so did the crew of Joe Jobe's plane, moving along on our left. Two of the bombs were short, but No. 3 hit the Jap at the waterline and No. 4 kicked her forty-five degrees to starboard, blowing most of her stern into the sky behind us. Her gunners were still firing at us as we jinked over to join Jobe, but she was done for.

Joe Jobe, meanwhile, had not been idle. Crossing the air strips with us at 150 feet, he had strafed a number of planes and started three major fires which were still violently blazing, and over the anchorage he had shot up a half-dozen or more coastal vessels while making his run on a cargo ship. At least five of the coastals and two valuable oilers were afire, and the cargo ship was ablaze from machine-gun bullets, when Joe spotted a larger ship farther out and broke his run to go for her. When we joined him a moment after our run on the destroyer, this second cargo vessel, a big one, was exploding all over the water. She kept exploding as we turned for home.

As we headed south toward Saipan, several ships were burning off the island's west shore and five distinct fires blazed on the island itself. We thought we had done a good job. Not a plane had been able to get off the ground to challenge us, and despite the vicious ground fire, neither of our planes was seriously damaged and not a man of either crew was scratched. But as I prepared to send a flash report to Saipan, battered Iwo Jima provided a dramatic postscript. Behind us, three separate and tremendous explosions lit the night, hurling flames and sparks high into the darkness. Only oil or ammunition dumps could have furnished the fuel for such a display. It was the pay-off.

"Skipper," said Shafe, "I can see it in the papers right now.

'Radio Tokyo reports Iwo Jima raided for first time by land-based bombers. Intrepid Japanese fighters turn enemy back with heavy losses. No damage to Japanese island. No comment from U. S. Navy.' "

About eleven o'clock, when our two returning planes were still a good many miles from Saipan, I saw a light winking on the horizon ahead. It shouldn't have been there, and it puzzled me. Saipan has no harbors of any size; our ships were lying in the open roadstead, vulnerable to attack by submarines known to be prowling in the vicinity. This mysterious light could easily be a sub signal of some sort.

It went out as I called Jim Park's attention to it. A moment later there were two of them.

For ten minutes these lights blinked on, went out, came on again. "Jim, we'd better report this by radio," I said. But then, as we neared the island, we saw that the lights slowly *descended* after they had blazed in the sky for a moment. "They aren't signal lights at all," Jim Park said. "They're flares."

He was right. On the northern part of the island, where hundreds of Japs were still hiding in caves and ravines, our Marines were sending up flares at regular intervals during the night to thwart any attempted suicide attack under cover of darkness. For our money, the star shells were excellent long-range lighthouses!

At Saipan we landed on the fighter strip in a drenching rain, and after I had reported to General Hale's Chief of Staff, Captain Davis of the Navy, Joe and I fumbled back through the dark to our camp. We had tents now, set up in the cane field near the air strip. They weren't much—the floors were still mud and the surrounding area was still littered with battle debris—but they were better than sleeping under the wings of the plane. Tired and soaking wet, Joe and I turned in about 3 A.M. and lay there listening to the drumming of the rain on the canvas.

Had we done a good job at Iwo Jima? I thought we had. Together we had seriously damaged and probably sunk a destroyer, a 6,000-ton merchant ship, two small tankers and three coastals, damaged a 3,000-ton merchant ship and set five coastals afire, destroyed at least 8 planes on the ground, probably destroyed 10 more and damaged many others, started 5 fires among shore installations and exploded an oil or ammunition dump. Two unescorted Liberators had done that.* I hoped Admiral Spruance would be pleased.

He was. In the morning, despite the Army's complaint that Isley Field was already overcrowded, the admiral ordered two more of our planes to Saipan, and I began to plan not only a return visit to Iwo Jima but a strike against the Bonins, as well.

23   *ONE CAME HOME ALONE*

EXCEPT for desperate enemy snipers who infested the island's hills and caves and would have to be pried loose one by one, Saipan was in American hands by July 10. Guam and Tinian, the other important islands in the Marianas group, obviously were next on the invasion schedule.

On July 16, two days after the Iwo Jima strike and while Jobe and I were still waiting for other planes of the squadron to arrive at Isley Field, I flew the Marine battalion commanders and intelligence officers of the Tinian landing force over that island's beaches for an hour or more, while they studied their objectives. The following day Hicks, Seabrook and Sheppard escorted Admiral King, Admiral Nimitz and Admiral Hoover to Saipan from Eniwetok. A day later, Janeshek and Bridgeman arrived with their crews, moved into our muddy little tent colony, and prepared to join us in the planned strike on the Bonins.

Information on the Bonins was meager. These islands lie about 120 miles north of Iwo Jima and are small and rugged—literally vol-

---

* Comment by Captain Taff: "The destruction caused by this strike emphasizes the value of a squadron trained for such tactics where the element of surprise and the aggressiveness of the pilots are used to the utmost. The employment of well trained and experienced squadrons in such a manner is recommended. Attention is invited to the short runway from which the flight started and the loading of the planes involved. Operations from this runway and with such a load are considered remarkable."

cano tops jutting from the sea. The three most important ones, Ani Jima, Chichi Jima and Haha Jima, roughly form an arrow pointing straight to Japan. Chichi Jima, only a little more than 500 miles from Tokyo, is the largest and is about four miles long and hook-shaped, the hook enclosing a harbor. (See map, p. 179.)

This was to be a dawn strike, just to keep the Japs guessing, and all four of us, Janeshek, Jobe, Bridgeman and I, were to take part in it. With the bomb load increased to 3,000 pounds and the Army still confidently expecting us to wind up in the mud at the end of the 3,800-foot runway, we took off in the dark and flew up to the north, fifty miles northwest of our objective, before turning. During the last hour of the flight the four planes were never more than 200 feet above the ocean, and when at last we slipped into the channel between Ani Jima and Chichi Jima, we were only 50 feet above the sea and hiding behind the mountains on our approach.

Again the Japs were taken quite by surprise. In a right echelon, with *Thunder Mug* leading, the four planes cleared the hills and came down over Chichi with machine guns in full voice. Over the neck of land at the northern end of the island we went, shooting at everything that looked like a target. Enemy gun positions took a drubbing as soon as they opened fire on us. A radio station was strafed and left burning. Barracks and other buildings were set afire as we swept out over the harbor. There, in addition to the usual cargo ships and coastal vessels, a number of Jap seaplanes sat like ducks on a mill pond.

One of these planes, caught in the act of taking off, suddenly became acrobatic as I bore down on it. Evidently the pilot was confused. He was in the air when *Thunder Mug* overhauled him, but instead of maneuvering to escape, he cut his gun and bounced on the water again. Then as our bullets bit into him he leap-frogged over the water, trying to "take off in all directions," and at last bounded up from the harbor to crash against the side of a hill beyond.

We too were clear of the harbor by then, and with Janeshek, Jobe and Bridgeman still holding the same carefully spaced formation, were on our way south. Together the four of us had sunk a cargo

BONIN ISLANDS

0 — 2
MILES

OTOTO
JIMA

ANI JIMA

OMURA

N

SUSAKI

CHICHI JIMA

To HAHA JIMA

ship, damaged a second one and seven coastals, set afire most of the important installations in the harbor area, and left the harbor itself littered with burning seaplanes. Now we turned across the southern end of the island and went for the airfield, strafing enemy planes on the ground as we passed.

How long had this taken? At 200 miles an hour you have traveled the length of a four-mile island in a little more than a minute, and you wonder how the boys on the guns can spot their targets at all. At treetop altitude, buildings, ships and parked planes rush so suddenly into your line of vision that they are almost blurred by the speed of the plane's passage. Now there is water under you, now a wooded hill in the midst of which an enemy AA gun blinks like a stop-light in the dark. Suddenly you are over a cluster of V-roofed buildings and rushing toward a row of planes parked wing-to-wing on an air strip.

All this time you are thundering along—and thundering is the word, for those four engines make a lot of noise!—just behind the probing fingers of fire from your own machine guns, the tracers spattering targets ahead and ricocheting in bright blobs of color. If time would stand still for a moment, you might be able to describe in some detail the results of such an attack. As it is, you do what damage you can and interrogate the crew later, piecing together then the bits of the picture as the boys try to recall what blew up and what didn't, what burned and what only smoked, and what it was that made that hellacious explosion to starboard as you jinked through the enemy's ack-ack.

We did enough damage on our run across the Chichi Jima airfield to cause the enemy at least a few headaches. When the four crews were questioned and cross-checked later, the score read 2 planes definitely destroyed, 11 probably destroyed, and at least 14 others damaged. Twenty-seven planes! Then we were over the sea again, still in formation, still unhurt, winging south across the twenty-mile span of water to Haha Jima.

There the attack had to be made in single file, for Haha Jima offers only a narrow space in which to work. But a lot of enemy war material was assembled in that slender corridor. In the harbor were six coastals; close inshore beyond the bulge of land were nine more.

Thunder Mug hit them first and set three of them on fire. Joe Jobe and Janeshek fired others. Bill Bridgeman made splinters of a good-sized cargo ship and gave the coastals still another drubbing. And then, with our bombs gone and ammunition boxes nearly empty, we called it a day and went home.*

This sortie to Chichi Jima and Haha Jima had kept us in the air fourteen hours, and by all the rules of the game we should have rested the following day, for pilots and crews were tired. It was important, however, to follow up the strike quickly, before the surprised Japs could puzzle out what had happened and reorganize their defenses. And because we had only four planes on Saipan, the rest of the squadron being still based at Eniwetok, we obviously could not turn the job over to anyone else. I discussed it with Janeshek, Jobe, Bridgeman, and some of the boys, and they agreed that we should go back to the Bonins the next day, with gas enough to hit Iwo Jima again, too. "Let's give 'em the works, Captain," they said. "All three islands at once!"

On the first take-off from Isley Field we had gingerly put aboard 2,000 pounds of bombs and 2,000 gallons of gas. On the mission to the Bonins we had increased the bomb load to 3,000, the gas to 2,200. This time there were 4,000 pounds of bombs on the racks, and the tanks held 2,400 gallons of fuel.

It was heartbreaking to have to fly such a load off a fighter strip when Army engineers just a short distance away were constructing an 8,500-foot runway for future use by B-29s. But there was nothing foolhardy in it. It was not a "stunt." We had always known our planes, and now, after two missions, we knew the field as well. This take-off, like the others, was accompanied by a pro-

---

* Comment by Captain Taff: "The highly commendable results of this action appear to have been obtained through a combination of excellent flying and marksmanship together with low-level tactics, aggressively and expertly carried out. As such, they are significant not only of a marked degree of skill on the part of pilots and crews, but also of a most laudable understanding of the desirability of surprise and a remarkable ability to use it to maximum advantage."

fuse amount of sweating, but with no assisting wind we still managed to get the big planes into the air.

Departing early in the afternoon, to reach the Bonins at dusk, we came in on Chichi Jima from the northeast this time, instead of the northwest. The four planes were in line abreast as we topped the hills. Once again the Japs were complacently sitting on their duffs, and once again, sweeping the harbor, we left burning ships and wreckage behind us.

The big score fell to Joe Jobe. His bombs ripped open a destroyer and blew her stern off with such an explosion that she was already settling when he turned to hit the airfield. There, however, he flew into heavy ground fire, and when we swung south to strike Haha Jima, his No. 3 engine was shot out, his bow and belly turrets were out of action with their gunners wounded, and he had holes in his wings through which you could have pushed a wash tub.

We were not aware of this until the four planes had swept over Haha, strafing coastal vessels in the harbor and bombing the buildings along shore. There the AA fire was intense and accurate, and Jobe's plane was hit again. As we pulled away toward Iwo Jima, he came up on the radio.

"I'd better drop out," he said, listing the damage he had sustained.

"Can you make it home, Joe?"

"I think so. I'll try."

If Joe thought he could make it back to Saipan, he probably would get there. He was that kind of pilot. Nevertheless, I worried. The weather was turning bad, the night was tar black. Joe's prospects of nursing a shot-up Liberator back over the 700-odd miles to Saipan on three engines did not look good. If I had known the full extent of the damage his plane had suffered, I would have worried even more. A .50-caliber gun had been kicked into his belly gunner's lap by enemy fire; most of his bow turret had been blown away; and it was only a miracle that Louis Paulunkonis, the bow gunner, had not been killed.

Joe dropped out of formation, jettisoned his shattered belly turret to lighten the plane, climbed slowly to 8,500 feet and started for Saipan on three engines. The rest of us went on to Iwo, planning to hit that island and pick up Jobe on the way back.

Iwo was by all odds the most important of these islands and the most heavily fortified. The shipping there, the number of operational aircraft on the ground, the powerful defensive system of light and heavy AA guns made it plain that the Japs were determined to hold the island, build it up, and use it to counter our seizure of Saipan. But again we caught the defenders napping and were on them with such speed that their vicious AA fire lacked accuracy.

We struck from the north, three abreast. Janeshek sailed down the coast, shooting up the boat yard and a number of small ships off shore. *Thunder Mug*, attacking inland, swung over the two air strips, dropping fifteen 100-pound bombs on hangars, barracks, planes and runways—which aroused the Japs to a frenzy and produced the most intense barrage of flak I had ever seen. Bridgeman, between us, went down along the western shore to the southern tip of the island, swung north again along the eastern shore, circled back over the air strips *after* I had alerted the defenses there, and flew his plane through the flak barrage while coolly pickling off his bombs. The maneuver should have turned his hair gray and given his crew heart failure, but it didn't. They came through it unscathed except for the usual flak holes, and picked us up six miles south of the island after blowing the best part of the Japs' major airfield to pieces.*

We turned for home, worried about Jobe and his boys. Flying through thickening weather over an invisible ocean, I tried several times to get Joe on the radio but received no answer. "Don't worry, Skipper," Jim Park said. "If anyone can make it, he can. He'll get home." But the hours went by and the radioman still glumly shook his head after every attempt. And then we sighted

---

* Comment by Captain Taff: "The report clearly conveys the gratifying results attendant upon a combination of superb flying, skillful tactics and excellent bombing and gunnery, a condition so characteristic and yet so singularly outstanding in the case of Commander Miller and in many of the crews of his squadron."

the flares over the northern tip of Saipan—the star shells we had thought were submarine signals—and out there in the darkness somewhere, Joe at last heard our signal. It was about two o'clock in the morning, raining hard. We had been out since one o'clock in the afternoon.

"I'm about out of gas," Joe reported. "Doesn't look as though we'll make it."

"Can you see the island?"

"I can't see a blamed thing. But Juelke (Ensign Oscar Juelke, his navigator) says we're on course."

"What about the wounded men, Joe?" I asked.

"They're all right. Neither is badly hurt. But we're still on three engines, my fuel gauges read zero now, and I don't know about my tires. We're shot up pretty badly on the port side."

I called the island, told them Jobe was in trouble and would need help. I gave them what I thought must be his approximate position. Then, maintaining radio contact with the crippled plane, we went on in.

A few minutes later Joe came up on the radio again. "Hey!" he shouted. "I think we'll make it after all! We can see those Marine flares!"

Janeshek and Bridgeman landed, and I radioed the field to alert them for Joe's arrival. The rain was a steady downpour now, and Joe would need space—and luck—to put his shot-up plane down on that little strip in the dark. I worried, too, about the 1,500-foot mountain just a few miles to the north; it was a real hazard. Circling the field, I anxiously searched the sky for Jobe's running-lights, while all of us did a little quiet praying in his behalf.

"Take it easy, Skipper," Shafe said.

Up there in the darkness with us was a P-61 fighter, its pilot eager to go down for a landing. I regret now that I lost my temper with that pilot. I was on edge, and he was undoubtedly just a kid. But he had certainly heard our talk with the tower and understood the situation, and when he complained that he had been out an hour and a half and simply *had* to go down because he was so tired, I got mad.

"Dammit," I yelled, "there's a shot-up Liberator coming in that's been out fourteen hours, and those boys are tired, too!" He stayed aloft.

Presently I saw Joe Jobe's plane, limping in with three engines and a feathered prop. She felt her way down through the rain, gingerly reaching for the field. She was okay. She was going to make it.

But as she touched, her port tire gave way, just as Joe had half expected. For an awful moment she slewed and skidded across the bumpy air strip. Then she pulled out of it and stopped. Safe!

I breathed again and stopped sweating, and old Shafe said quietly, "Are you all right, Skipper?" He knew I was all right—then. He understood. When you have sweated your boys home on mission after mission, you eventually reach a point of not caring a damn who sees the tears in your eyes. The fact that we had just completed our third successful attack on the enemy's inner-fortress islands in six days was less important—far less important—than knowing that Joe Jobe and his crew were home safe.

[*The official score for these three strikes at the Japanese inner-fortress islands stands as follows:*

Destroyed or Probably Detsroyed: 2 destroyers, 2 large merchant ships, 2 small tankers, 3 coastal vessels, 33 planes.

Damaged: 3 cargo ships, 20 coastal vessels, 20-30 planes, 1 destroyer escort.

Barracks, hangars, oil and ammunition dumps, and many other installations were left in ruins.—H.B.C.]

24   *THE MUG'S LAST FLIGHT*

CONCLUDING his interrogation, Lieutenant Ted Steele, the squadron ACI Officer, scowled at Delahoussaye's trouser leg and said, "What happened to you?" The pants looked as though they had been pulled through a barbed-wire fence.

Delahoussaye grinned. "Nothing much, Mr. Steele. I was shooting at the target when all at once this Jap tracer bullet zipped through the deck and went up my pants leg." He laid open the charred cloth to show that his own leg, beneath, was more or less unharmed. "It went out at the knee without doing any damage. I got burned a little, that's all, when my pants caught fire."

"Why didn't you douse the fire with water from your canteen?"

"I didn't have a canteen, Mr. Steele. I forgot to carry one this trip."

"You ought to carry one."

"All right, Mr. Steele," said Delahoussaye, "I will."

A day or so later the crew of *Thunder Mug* appeared again before Ted for interrogation after a mission. This time Delahoussaye didn't wait for Ted to get around to him. He solemnly approached the ACIO's desk and placed a jagged piece of metal under Ted's nose.

"What's that?" Ted asked.

"My canteen," said Delahoussaye. "I carried one like you told me to. I had it on my hip. This is what the Japs did to it."

Ted gingerly picked up the battered bit of metal. "Delahoussaye, don't you think you'd better stay on the ground the next time your crew goes out? Two close ones like this, one on top of the other—you're pressing your luck. Get yourself a sore throat or a headache for a few days." He put the remains of the canteen down again. "I mean it. I'm serious."

But Delahoussaye was grinning, as usual. "Hell, Mr. Steele," he said, "they can't get me. Those Japs can't spell my name." At the door of the Quonset he paused. "Even *you* can't spell my name, Mr. Steele!" he declared, and went out chuckling.

"You should talk," piped up Jaskiewicz. "*Nobody* can spell mine!"

I sometimes tell that story to people who ask how the crew of *Thunder Mug* happened to run up such a record. And then I tell them what happened in Truk, on a mission we made July 29 with Commander Don Gumz.

After the three strikes at the Bonins and Kazans, I had returned to our main base at Eniwetok, and Hicks, Bundy, Seabrook and Kasperson had gone up to Saipan to take the next trick there. At Eniwetok, VB-108 had been relieved by a new Liberator squadron, and it was up to me to pass on to Don Gumz, the new skipper, some of the dope we had picked up in our months of combat—just as VB-108 had given us the dope when we were new. Don was no novice at combat, of course. In another type plane he had been out and done plenty. My job was to show him around his new stamping ground.

There had been reports of enemy shipping in Truk, and on the previous night I had gone in there to investigate. Despite our many raids on the atoll, the enemy had again been taken by surprise. His ground defenses were weak, he put no planes in the air to attack us, and in a leisurely fifteen-minute tour of the islands I had sunk a cargo ship and a number of landing craft.

The Truk defenses obviously were much weaker than we had found them on earlier raids. Still, there *might* be some important shipping hidden among the islands as reported. So Don Gumz and I went calling.

Our two planes hit Truk at dawn, approaching leisurely from the northwest, in and out of thunderstorms, to kill time while waiting for the light to improve. Even with daylight, however, the islands were deceptive, and I made a run on a ship only to discover, when almost on it, that it wasn't a ship at all but a stripped island that looked like one. Probably in reporting enemy vessels in Truk Lagoon, Army pilots at high altitude had mistaken some of these small, ship-shaped islands for vessels. It was an easy error to make, and the Japs had furthered the illusion by removing the trees or tops of trees from many of these islands, perhaps to fool us into wasting our bombs on them.

Our plan of attack was simple. Don was to circle the north side of Moen, come down east of Dublon and join me as I completed a tour of the main anchorage and flew through the Dublon-Fefan corridor. Well inside the lagoon we parted company to go our separate ways.

The anchorage was empty. I swung south to look in on Mesegon, and found three small coastal craft which the boys shot up with their guns, but the Mesegon air strip was still not operational and I turned toward Uligar Pass. Suddenly Whisky Jazz Jaskiewicz called delightedly over the interphone: "Hey! Here comes Commander Gumz with an escort!"

Commander Gumz indeed had an escort. He had five Jap fighters on his tail! Here is his own version of what he saw when he came tearing across Truk Lagoon to join us:

"When the Zekes and Oscars lit after me, I hightailed for you to get the added protection of your guns. You were cruising along, evidently looking for us and hunting for some targets. Jaskiewicz was lazily leaning on his folded arms in the open waist hatch.

"Well, he heard us coming and looked up. Didn't budge off his elbows—just watched us as we came jinking up with those five Japs making passes at us. You'd have thought he was back home, gazing out a window. Until we were in range. Then, without a wasted motion, he reached out for his gun, swung it around and began shooting."

That's the kind of crew I had. It just happened to be Jaskiewicz on this occasion; any other member of the crew might have been equally cool. In the face of danger they invariably displayed a quiet dignity that made them great.

[*But according to the officers and men of the squadron, much of this quiet dignity stemmed from Bus Miller himself. Here, for instance, is a story Lieutenant Steele tells about the commander:*

"*The Skipper took* Thunder Mug *into Truk one day on a routine search, and I went along because it was easier to ride with him than to interrogate him afterward. When we began our bombing run on one of the ships in the pass between Dublon and Eten, the Skipper was eating a sandwich. I stood behind him, between the armored backs of the pilot's and co-pilot's seats, watching the curtain of flak that rose in front of us—flak so thick that you could have stepped out of the plane and walked on it. Scared? I was petrified. But Commander Miller went right on eating that sandwich—not just nibbling at it, but biting big comfortable chunks*

out of it and swallowing them. And then, turning to me, he said
solemnly, 'Ted, these are wonderful planes. Did you ever stop to
think what makes 'em fly the way they do?'

"Perhaps that was an act deliberately put on to show the crew
that their captain, who after all held their lives in his hands, was not
afraid. But I don't think so. I think it was simply the Skipper ex-
pressing the confidence he felt in his plane, his crew, and himself.

"But if it was an act, it worked admirably. After the boys had
flown with Commander Miller for a while, they began to talk and
act as he did. And that is part, at least, of what made them great."—
H.B.C.]

Enemy fighters were not exactly rare in Truk. On almost every
occasion they had risen to attack our carrier planes and Army
bombers. We, however, didn't usually flush them—we came in so
low and worked the element of surprise so carefully that we were
able to pin them down with bombs and machine-gun fire before
they could get off the ground to challenge us. Janeshek, "thumb-
ing his nose" over Truk in broad daylight, had once been chased
by five fighters, but most of the time we were concerned only with
the enemy's ground fire or fire from the ships we attacked.

Now, with five Zekes and Oscars stabbing at us, we retired at
best possible speed, which wasn't much because both planes were
low on gas and still toting bomb loads, and we had to stay in
"cruising combination" on our engines. The Japs, of course, stayed
with us.

I turned for a rain squall about fifteen miles to the east and ordered the boys to pull up the belly turret to cut down our wind
resistance. Smarter Japs would have taken advantage of that, coming up from below where we were then vulnerable. These didn't.
As we headed for the clouds, the fighters kept darting at us, making slight overhead passes and flat side runs. Thwarted by the com-
bined fire-power of our guns, they appeared to be timid, uncertain, and preferred to stay just out of range.

Presently one of them dropped out, smoking from a hit, and another, attacking from ahead, dropped phosphorus bombs—umbrella-

like affairs with long white streamers that floated weirdly past *Thunder Mug's* wing tips as I jinked. In and out of the clouds we played tag with them, holding our fire until they came in, then watching them falter and turn away as the Liberators' guns discouraged them.

A second Jap began smoking and peeled away, but the other three still heckled us as we plugged along, leading them farther and farther from their base at Truk. At last after thirty-two minutes they had had enough, and we turned for Ponape to complete the tour of our search sector and hunt up a target for our bombs.

Ponape was nothing like the old days, when Jokaj Joe had given us so many royal receptions. The newer air strip was still not in operation; the old one was empty of planes; the town appeared deserted. We caught a little light and medium flak from the ground defenses but found no worthwhile targets, and so, with our bombs still aboard and the boys yawning at their guns, I swung down over the small islands in the harbor.

There we stirred up some action. The Japs had a seaplane base in the harbor and were evidently hopeful of keeping it a going concern. The AA fire was heavy and accurate as our two planes punished the place with machine guns. As we turned for home, I realized that *Thunder Mug* had been hit. In retaliation for the damage done by my seven 1,000-pound bombs—not to mention those dropped by Don Gumz—the Ponape Nips had sent up some all-too-accurate flak, and we had caught a sizeable portion of it.

This time it was moderately serious. The flak-torn lines of the hydraulic system began to leak precious fluid. The radio fizzled out. I began to worry about the brakes.

Don Gumz arrived at Eniwetok first and went on in. In *Thunder Mug* I doped along while the crew checked and tried to repair our damage. Even Paul Ramsey was stumped this time, however. "Our hydraulic fluid's gone," he reported, shaking his head. "That's bad, Captain."

It was bad enough. The air strip at Eniwetok had not been designed to accommodate brakeless Liberators. I tested the brakes gingerly not wanting to use what little fluid might be left. They appeared to be working. But we couldn't be sure.

As we neared the island, Ramsey came up into the cockpit and said, "Skipper, why don't we rig parachutes in the waist hatches and trip them as we land, just to be on the safe side? They ought to slow us down, at least."

"All right, go ahead," I agreed. "And tell the rest of the boys to run aft as we go in. That will hike her nose up and dig the tail skag into the ground."

The parachutes were rigged as I circled the field, pumping the flaps and landing gear down by hand. I couldn't radio the tower that we were coming in crippled; our radio had been shot up beyond repair. So we droned down over the end of the island, hoping for the best.

I reached for the interphone. "All set back aft?"

"All set. Just give the word."

The wheels touched the coral. I applied the brakes. They wouldn't hold us, and my shout over the interphone, "Trip chutes!" was a waste of time. The interphone picked that moment of all moments to cease functioning, and the boys never heard the word.

At something like 60 miles an hour I tried to ground-loop *Thunder Mug* as she sped toward a row of parked carrier planes at the end of the field. We missed the parked planes, but the ground-loop fizzled. We slewed off the field, traveled down across the road, came within inches of colliding with a jeep, and plunged down the embankment fifteen feet into three feet of water, smashing the bow turret (which was unoccupied, of course) and burying *Thunder Mug's* nose in the coral. For just a moment we all sat there—dazed, I think, by the realization that the plane which had carried us all over the Pacific for nearly eight months would never again roll down a runway.

Jack Simmen, the bow-turret gunner, climbed out of a waist hatch and waded out to his turret, which had broken off and dropped into the water. Simmen was the workhorse of the crew, a big, quiet boy who had spent nearly all his time at the plane, grinding away at the guns, patching up holes, cleaning her up. Waist-deep in the ocean, he put his arms around the shattered turret and stood there holding it. His gesture was a caress. He spoke, I know, for all of us—though he said nothing at all.

The following day we watched in silence while the Seabees pulled old *Thunder Mug* out of the ocean with bulldozers and towed her to the boneyard.

25   *KAS*

AUGUST 2 was the anniversary of our commissioning, but the squadron could not be together for a celebration. Some of us were on Eniwetok, others were living in the mud on Saipan. Nevertheless, we did our best to mark the occasion.

At Eniwetok, the new squadron generously assumed our patrols for the day, and with the help of an extra case or two of beer we held an outdoor party. It wound up with the enlisted men tossing their officers into the ocean—an old Navy custom on such occasions —and with me running a marathon around the Quonsets with half the squadron personnel in hot pursuit.

On Saipan, where living conditions were more primitive, the four crews of our advanced echelon held a celebration in a mudhole. The usual rain had been falling, the tents were inundated, there was no place in camp to sit, or even to hang anything up. Undeterred, the boys caught two small Japanese pigs. They were too chicken-hearted to kill the pigs; Tom Seabrook, our "Baedeker of the South Seas," had to do it for them. They roasted the animals on a spit made from a discarded wagon axle, devoured them with green onions from a Jap garden, and washed them down with hot canned beer, after which they sat around in the mud and rain until three o'clock in the morning, singing songs about the Navy, the Marines, the Army Air Corps (!), and especially of home.

One of those who celebrated our anniversary with enthusiasm was Lieutenant Elmer Kasperson, who had been with the squadron since its commissioning. Always quiet and cheerful, Kas had unlimited energy, a rich, deep sense of humor, and a genuine love of flying. He liked to explore the Pacific in his Liberator. He liked to discover new islands. His crew was always coming home with stories.

For example: Kas had reached the end of his search sector one day and was about to turn for home when he saw an enemy destroyer ploughing through the sea up ahead. The Japs on the destroyer saw the Liberator at about the same time, and confidently manned their guns. As Kas went in to attack, they laid a solid screen of flak in front of him.

Kas turned away and went in again, feinting, but the Japs would not be fooled. Again, and a third time, the ship's gunners hung an accurate curtain of fire in the plane's path.

Kas drew away. "We wouldn't have a prayer going in there," he admitted to Ensign Hindenlang, his co-pilot. And to his radioman: "Tell 'em we'll be back some other day."

The radioman blinked a message. "Okay, you guys, you win this time but we'll be back."

Promptly the destroyer blinked an answer in English. "Okay yourself. Roger."

Another time Kas was coming out of Truk when he saw what looked like an enemy cargo vessel. He made a run on it, discovered it was a hospital ship, veered away without attacking, and blinked, "How's business?"

"Okay," the Jap answered. "Fine."

Kas was always coming home with stories like these. But his crew had to tell them—Kas wouldn't.

Early in August, on Saipan, Hicks and Bundy and Seabrook and Kasperson undertook the difficult and dangerous task of heckling the Kazans and Bonins between two days of attacks by carrier-based aircraft. The carrier planes were to hit Iwo Jima and Chichi Jima during the daylight hours of August 4 and 5. The four Saipan-based Liberators, operating in relays, were to pester these islands throughout the intervening night, denying the enemy an opportunity to attack our carriers, repair his damaged airfields, or bring in new planes.

Lieutenant Commander Hicks and Lieutenant Commander

Bundy drew the Iwo Jima assignment, and Hicks went first, taking off from Saipan at seven-thirty in the evening. He found Iwo Jima hidden under a storm that reached 20,000 feet into the sky. With zero visibility he heckled the island with small bombs, crossing back and forth over it through the storm, until two o'clock in the morning. The enemy ground fire was blind and ineffective. The Japs were shooting at a ghost plane.

Bundy, flying with my own *Thunder Mug* crew, arrived over the island as Hicks departed. The weather was still atrocious, the island still hidden by squalls and clouds. Continuing the heckling, Bundy flew back and forth over the airfields as Hicks had done before him.

My crew told me later that they had never seen a pilot fly so audaciously low under zero-zero conditions as Bundy flew that night. "We all but scraped the tops off the trees," they said, marveling at Bundy's adroit handling of the plane, "and there were times when we thought we never would pull up over that bluff at the northern end of Iwo."

But Bundy was unlucky. One of the blind-firing ground guns picked him up. Enemy AA exploded among the plane's oxygen bottles and set the ammunition boxes on fire, holed the rudder, damaged an engine. He had to fly home from Iwo Jima to Saipan, through incredibly bad weather, on three engines— "And, Skipper," my crew said, "he handled that plane as if he'd built her with his own ten fingers." At seven in the morning he put the crippled plane down on Isley Field, with all hands safe.

Lieutenant Seabrook and Lieutenant Kasperson, meanwhile, had drawn the Chichi Jima assignment and planned to heckle the airfield and island together. Kasperson took off first, at 7:10 in the evening, just before Hicks left for Iwo. Hicks passed him out over the ocean, half way to the target, and saw the tracers as Kasperson's gunners test-fired their weapons. Seabrook departed from Saipan a little while later.

Seabrook arrived at Chichi soon after midnight and found the same ugly weather that was making the task so difficult at Iwo. He heckled the place, pinning the Japs down, until quarter past two. Then, concerned because he had seen no sign of Kasperson, he returned to Saipan.

Old Kas, who loved to fly and always took his sense of humor along with him, did not come back that night. He may never have reached the target. He may have reached it and been shot down, or flown into one of Chichi Jima's rugged hills while trying to get his bearings in the storm. Kas and his crew may have perished; they may be prisoners. We don't know.

26   *"WELL DONE!"*

THE loss of Kasperson and his crew was a numbing blow to the squadron, doubly tragic because it came on what was destined to be our last important striking mission. During the remaining few days of our tour of duty we flew a few missions to Wake, Ponape and Truk, but the major work was finished. We had been out long enough. On August 7 I went out to Admiral Hoover's ship to consult with him, and the admiral said he was requesting permission to send us home.

It was good news. We were tired, and all our planes were in need of major overhauls. I returned to our Eniwetok camp to pass the word to all who were there—some were on patrol—and found most of them at the movies.

These were open-air movies, down by the ocean. The boys were sitting there in a drizzle, doggedly soaking up their "entertainment." I climbed up on the platform in front of the screen and told them the news, and the movies—for the men of VB-109, at least—were suddenly of minor importance. The boys stood up and cheered. Then I boarded a plane of another squadron that was taking off to relieve our planes at Saipan, for I wanted very much to be the first to deliver the good news there, as well. Our crews on Saipan had been hunting for Kasperson night and day since the night of his disappearance. (But despite a thoroughly organized search, in which PBMs, PBYs, carrier planes, submarines and even a destroyer had participated, nothing had been learned.)

Luck was against me. Arriving at Saipan, I found that Admiral Spruance had already sent our planes back to Eniwetok—Admiral Hoover had ordered the new squadron to take over—and the only 109 men on the island were Ted Steele, our ACI officer, and John Sladich, a yeoman. They were lonely, both of them—sick of the Saipan mud, the flies, the mosquitoes and the incessant rain. The loss of Kasperson had sobered all of us.

I was no more eager to remain on Saipan than they were, and so I sent a dispatch requesting the plane on patrol in that sector to stop by and pick us up. But while we waited in the mud, cussing out our luck, the weather worsened and the rain began to fall in bucketfuls. Then a fog thickened along the coast, moving in to blanket the mountains and hide the airfield, and we knew there was no use waiting for the plane because the plane wasn't coming.

It was Ted's birthday, and there was simply nothing else for us to do on that ugly little island, while waiting for transportation back to Eniwetok, but make the most of the occasion. Even on Saipan there was bound to be a bottle of something somewhere. We went on the prowl and came up, just before dark, with a pint of whisky.

Unhappily, however, there was but one comfortable place in the camp area in which to sit with a bottle of whisky—or even just to sit—and that, at the moment, was being occupied by Commander Don Gumz and Lieutenant Charles Cervone of the new squadron. They were scheduled to fly the next day and needed their sleep. So, to be sure that we would not annoy them with our talking, we moved out of this cozy little tumbledown shack and hunted up other accommodations.

Outside the shack a jeep was parked, with a canvas cover stretched over it to keep out the rain. Beneath this canvas we celebrated Ted's birthday, passing the bottle back and forth now and then to build up a bit of resistance against the damp night air. We talked mostly of home, of Ted's wife, Peg, and the baby that was coming (Ted might learn a lot from a five-time Papa like me!) and of Thelma and the youngsters in Jacksonville, and of what it would mean to us to be home again, after so many months of living in a strictly masculine world on small, crowded Pacific islands.

In the midst of this quiet reminiscing, Ted suddenly raised a hand. "Listen!" he said. I heard a soft clicking sound that puzzled me. "Must be Commander Gumz," Ted whispered, "tapping his Academy ring on the iron hinges of his cot, asking us in a nice way to be more quiet out here. We'd better move." So we abandoned our nest under the canvas and went and sat in the cane field, where the ground was sodden and the rain watered the whisky.

It was eerie, sitting there in the dark. About a mile and a half away, Marines were scratching with machine guns and flame-throwers at dug-in Japs who had refused to surrender, and in the cane field itself, other Marines were hunting down Jap snipers. Every now and then a rifle spoke, a bullet whistled, and we heard voices. Then the black sky would open, dumping gallons of rain on us, and we would scurry back to the shelter of the canvas-covered jeep. And when we had been there a while, trying to continue our reminiscing without speaking above a whisper, that same queer clicking sound—Commander Gumz tapping his ring on the cot to quiet us—would send us back to the cane field. This went on until at last the bottle was empty and we crawled into bed, wet through and shivering, but feeling better, as men will, for having spent our last night on that bloody island in quiet memories of home and wives and children.

In the morning I apologized to Don Gumz for having kept him awake, and he looked at me in bewilderment. "Awake?" he said. "I slept like a log."

"But the clicking noise," I protested. "*That* clicking noise!" And we both looked at the rafters overhead, where something was again making a sound like a ring tapping on metal. And then we laughed, and I felt better, knowing we had not kept the commander awake after all. But the little brown chameleon sitting up there on the rafter, snapping its tail or its scales or whatever it is that chameleons snap to produce that clicking sound, was entitled to the last laugh, after all.

Dear old Saipan. That day, when Ted, Sladich and I boarded a plane as passengers for the trip back to Eniwetok, our troubles began again. The field was jammed with fighter planes. For an hour and a half we taxied around them and through them, trying to find room to get to the runway for a take-off, and finally, after sinking wheel-deep in the mud at the edge of the strip, we had to be hauled out by tractor and bulldozer. Away at last, with someone else at the controls for a change, I stretched out with Ted on a mat of bedding in the waist compartment, and with nothing to do and hours to do it in, we whiled away the time remembering old songs.

"Ted, I used to play trombone in the church choir back in Winston-Salem. That was a long time ago."

"Skipper, I'm thirty-two myself, in spite of this shimmering red hair. It will be good to get home. Do you suppose my baby will have red hair?"

We sang *Whispering* and *My Blue Heaven* and *Side by Side* and *Who* and *Smoke Gets In Your Eyes*. We hummed hymns and snatches of opera and bits of symphonies—just as I had used to do while riding in the evening along North Carolina roads with Thelma Davis. But now the road was a sky road, and the high-school jalopy had become a four-engined Liberator, and beneath us lay a hostile ocean over which we had flown thousands of hours of patrol. And as night came on, we lay there watching the stars out of the waist hatches . . . still singing old songs and talking of home.

At Eniwetok we were given the red light and ordered to proceed to Engebi, at the north side of the atoll. It was midnight when we landed, and so we spent the night in the open after all, under the wing of the plane at the edge of the Engebi air strip * and went on to Eniwetok in the morning.

* *"This didn't bother the Skipper much," reports Ted Steele. "Dead tired, we curled up in blankets beneath the plane and were asleep instantly. About 3 A.M., however, a squadron of fighters took off, growling down the runway about a yard from our heads. At 4 A.M. a squadron of Army bombers thundered by, and at 5, a squadron of Liberators. At this point, sleepily opening his eyes, the Skipper frowned at me and said, 'You know, Ted, we must be somewhere near an airfield, huh?'"—H.B.C.*

VB-109 flew its last patrol on August 16, and two days later was relieved. In seven and a half months of operation in the combat zone, we had made 1,141 operational flights for a total of 12,153 hours of flying time. Now, with three messages of farewell from high commands to all hands, we were ordered to return to San Diego for re-organization and re-assignment.

From Admiral Hoover, Commander Forward Area, Central Pacific: "In your excellent performance of duty in the Forward Area, Central Pacific, you have set a high standard for those who follow. Well done!"

From Admiral Towers, Commander Air Force, U. S. Pacific Fleet: "Commander Air Force, Pacific, congratulates the officers and men of Bombing Squadron 109 on completion of a successful and most effective tour of duty in the combat area. Your record is outstanding and performed in conformance with the highest standards of the Naval Service. Well Done!"

And from Admiral Chester W. Nimitz, Commander-in-Chief, U. S. Pacific Fleet: "It is enough to say that the enemy will be glad you have left the forward area. Congratulations on an outstanding tour of combat duty. Well Done!"

27   *GOOD TO BE HOME*

"NOTICE to all Patrol Plane Commanders and Plane Captains: VB-109 planes are to be taken back in combatant operational condition. No turrets, armor plate or guns are to be removed. Ammunition and bombs are the only ordnance gear that shall be taken out before leaving this base (Eniwetok)."

So, then, we were to go home the way we had come out—in our own planes. Not all of the original fifteen planes of the squadron would be making the eastward journey. Three, with their crews, were missing in action, and Hale Fisher, our shipmate, was gone. *Thunder Mug* and two other planes had cracked up and were rest-ing in Pacific island boneyards. One had been destroyed on the ground by enemy bombs.

Some of these lost planes had been replaced during the tour, however, and the homeward flight to the Naval Air Station at Kaneohe, Hawaii, was made by eleven Liberators. The extra pilots and crews went along as passengers.

They were tired Liberators. Nearly all of them bore many scars made by enemy bullets and flak, and had been patched again and again. The once buxom damsels on the fuselages of some looked more like hard-bitten professionals. Paint was peeling from proud names under the cockpit bubbles.

Nevertheless, one by one the squadron's planes arrived at Kaneohe and one by one they took off on the final leg of the journey to San Diego. Last to leave was *Flying Circus*, the battle-battered Liberator in which I was flying with my old *Thunder Mug* crew.

She had been Lieutenant Commander Bundy's plane and she had been a brave lady. In her, Bundy and his crew had flown many thousands of combat miles and rolled up an impressive record. But over Iwo Jima, just before the end of our tour, she had been shot up badly, and now, despite the work done on her tired old engines, she was wheezy and temperamental. When I took off from Kaneohe I had my doubts that we would get to the West Coast without some sweating.

We didn't. A few hours out, one of those weary engines gave up the struggle and back we limped to Kaneohe for repairs. Unhappily the boys rolled up their sleeves.

We tried again. This time, after more than six hours in flight, we thought we were going to make it. Cheerfully we discussed our plans for the immediate future—a future to be filled with phone calls, train reservations, and quick trips to homes scattered all over the country. We were almost "over the hump." Then an engine conked out, and back we went over 1,080 miles of ocean, on three engines, with the belly turret and every other movable object dropped overboard to lighten the load.

"We'll try once more," I said.

"Skipper, we're jinxed. We've ridden our luck too long."

They meant it, too. But we had *never* missed a combat mission, and only once, on a Wake Island raid, had we had to return to base to pick up another plane in order to complete a mission once started. Oddly enough, our nemesis then had been this same plane of Bundy's.

"Once more," I insisted.

We tried once more. Before we had been out an hour, the plane gave up the struggle and I had to take her in for a landing with a full load of gas aboard.

That was the end of our patience. But for the kindness of Admiral John Dale Price, who had made our stay at Kaneohe a thing to be remembered, we probably would have abandoned the battle before. Now we alone of the entire squadron were still marking time at Kaneohe, while the rest of the boys were at home or on their way. It was too much.

While I went to the chief of staff for permission to take another plane, the crew found a pot of paint and made a few alterations on Bundy's old Liberator. Her number, 48, was changed to read "Golden Gate by '48." Her handsome damsel was given a beauty treatment in reverse. Her name was changed from *Flying Circus* to *The Crock*. Poor Bundy! But the next time we took off from Kaneohe, in a modern Liberator of the "56" series, we made it. The U. S. A. at last!

I hurried to a telephone and said what thousands of men must have said before me, and thousands more will say after me. "Darling, I'm back. I'll be home." But there was some official running around to do first—Washington, New York, Washington again—and it seemed a very long time later when at last I stepped off the train in Jacksonville.

It was late, nearly midnight. The station was jammed with servicemen from near-by camps, waiting for trains to take them north. Outside, a lone cab stood at the curb, with two or three passengers already aboard and waiting for the driver to fill his vehicle—for these were days of "cab pooling" and you rode a taxi as you would a bus. I put down my bag.

"Are you going out by Shirley Avenue?"

"Sure," the driver said. "Get in. I'll be leaving in five minutes."

I hadn't telephoned. The train might be late and I had not wanted her to wait in the station at that hour of night. Now I sat in the cab—five minutes, ten, fifteen. Impatient.

"Driver, when do we get started?"

"In five minutes, I told you."

"Dammit, that was fifteen minutes ago!"

"So what? I have to fill up my cab, don't I?"

Ten minutes more. I wasn't used to these home-front tribulations. In a Liberator you travel far and fast, and I had come nearly half way around the world, only to be held up at the last mile. I got out, took my bag and began walking. Then, surprisingly, another cab drew up alongside me on the empty street.

There was a light in the sitting-room of the house on Shirley Avenue. Thelma, of course—probably reading a book. The kids would be asleep at this hour. I went around to the back door and fumbled for a key I had not used in months. Opened the door softly. Tiptoed into the kitchen.

She was talking on the telephone in the sitting-room. Talking quietly. I could hear the voice but not the words. So I waited. Just a little while. Put my bag down.

"Hey," I said at last, "what does a guy have to do to get a welcome around here?"

I heard a little gasp. Then, into the phone: "He's home! He's here now!" The phone clicked and she was running.

It was good to be home.

28   DEDICATION

PROBABLY this is an odd place for a dedication. But if you have carried on this far, perhaps you will go a little farther. I hope so, for there are some things I want very much to say, not only about the officers and men of Bombing Squadron 109 but about the pilots and crews and ground personnel of all other search squadrons of the Navy—whether they fly Liberators, Catalinas, or Mariners.

In the beginning, the Navy's Liberator squadrons were classified as "bombing" units, a designation decided upon because the four-engined Liberator is primarily a bombing plane. Our job was not bombing, however. It was to act as the long-distance eyes of our combat forces—to range far from our advance bases, searching out the enemy and reporting his movements, so that our land, sea, and air combat units would be "in the know" always. (Remember Ensign Jewell Reid, who flashed word of the approaching Japanese fleet in the Battle of Midway, which has been called the turning point of the war? He was flying a PBY patrol plane, 700 miles from base.)

Today we are classified as a "search, reconnaissance and photographic" force, which is a change in name only. Our primary job remains the same. We still fly long, monotonous, dragged-out hours over hostile waters and enemy-occupied territory, almost always alone, relying solely on our own armament for defense against whatever the enemy may send against us.

So, then, your Navy search plane is a long-range pair of eyes with a radio set, flying alone, far from home, over an enemy ocean, to find out what the enemy is up to. That is job No. 1.

Job No. 2 was conceived by the men themselves. They were flying daily over important targets of opportunity—enemy shipping on the high seas, ships in little island harbors, radio and weather stations, airfields, barracks, and assorted military installations. Why not, they argued, hit these targets whenever the chance presented itself?

The Navy replied, "Why not? Aggressiveness, as always, is encouraged. Provided his search sector is properly covered, any Patrol Plane Commander who likes to step out of his role as search plane pilot and do a little 'going to town' on the Japs is encouraged to do so."

And so the Navy's search planes began brightening their long, monotonous patrols with attacks upon the enemy. On their own initiative they went "out and beyond" their assigned search sectors to "hit the enemy when we can, as often as we can, and wherever we can find him." They ran up formidable scores.

But remember this: In almost every instance, these attacks were made by lone planes which had already been out for hours on patrol and were a long way from home when they went Jap hunting. They were opening the cage door to spit in the tiger's eye.

They are still doing it. You will be hearing more from them, much more, before the war is over. Some, like Elmer Kasperson, Sammy Coleman, Johnny Herron and their crews, may not come back—but all of them will always be remembered.

To them all, and especially to the officers and men of my own squadron, this book is dedicated with heartfelt pride and respect.

BOMBING SQUADRON 109

Cmdr. Norman M. Miller, USN, Pilot, PPC, Commanding Officer, Winston-Salem, N. C.

Ens. James R. Park, USNR, Co-Pilot, Los Angeles, Calif.

Lt. (jg) Robert K. Schaffer, USNR, Navigator, Saginaw, Mich.

Paul K. Ramsey, ACMM, USN, Plane Captain, Air Gunner, York, Penna.

William E. Fitzgerald, ACOM, USNR, Gunner, Ordnance Chief, Brooklyn, N. Y.

Adron G. Whitson, ACRM, USNR, Leading Chief Radioman, Blanchard, Okla.

Wayne Young, ARM1c, USN, Radioman, Air Gunner, Seattle, Wash.

Henry C. McNatt, AMM1c, USNR, Air Gunner, Mech., Compton, Calif.

Edwin L. Dorris, ARM2c, USNR, Air Gunner, Mech., Kansas City, Mo.

Robert Gariel, AMM2c, USNR, Air Gunner, Mech., San Antonio, Tex.

Thomas W. Delahoussaye, AMM2c, USNR, Air Gunner, Mech., Lafayette, La.

Lawrence B. Johnson, AOM2c, USNR, Air Gunner, Ordnanceman, Grampian, Penna.

Bernard R. Jaskiewicz, AOM2c, USNR, Air Gunner, Ordnanceman, Bay City, Mich.

Jack A. Simmen, AMM3c, USNR, Air Gunner, Mech., no address

Crew No. 2

Lt. John E. Stewart, USN, Pilot, PPC, Sharonville, Ohio

Lt. (jg) Eugene A. Terrell, USNR, Co-Pilot, Hannibal, Mo.

Ens. William L. Rarick, USNR, Navigator, Eaton, Ind.

Marwood E. Ebert, AMM1c, USNR, Plane Captain, Air Gunner, Milbank, S. D.

Jorma O. Allonen, ARM2c, USNR, Radioman, Air Gunner, E. Weymouth, Mass.

Ovey Janies, AOM2c, USNR, Air Gunner, Ordnanceman, Mamou, La.

Miles W. Morris, AMM2c, USN, Air Gunner, Mech., Marlin, Tex.

Orval F. Sanderson, AOM2c, USNR, Air Gunner, Ordnanceman, Portland, Ore.

Walter R. Stilson, ARM2c, USNR, Radioman, Air Gunner, Franklin, N. Y.

Jack C. Wilhite, AOM3c, USNR, Air Gunner, Ordnanceman, Oklahoma City, Okla.

Richard S. Levy, AMM3c, USNR Air Gunner, Mech., Sanford, Fla.

Crew No. 3 "MISSION BELLE"

Lt. Thomas Seabrook, USN, Pilot, PPC, Navigation Officer, Ridgewood, N. J.
Lt. (jg) Clarence W. McKee, USN, Co-Pilot, Phoenix, Ariz.
Ens. Fred H. King, USNR, Navigator, Fort Wayne, Ind.
Arthur E. Babcock, AMM1c, USNR, Plane Captain, Air Gunner, Adrian, Mich.
George E. Holden, AMM1c, USNR, Air Gunner, Mech., Burley, Idaho
Donald E. Houghton, AOM2c, USNR, Air Gunner, Ordnanceman, Perry, N. Y.
Floyd D. King, AOM2c, USNR, Air Gunner, Ordnanceman, Newark, Ohio
Thomas J. Burns, ARM2c, USN, Radioman, Air Gunner, Long Island, N. Y.
Norman A. Bye, ARM2c, USNR, Radioman, Air Gunner, Lily, S. D.
Bill R. Haas, AMM3c, USNR, Air Gunner, Mech., Fort Gibson, Okla.
William H. Enid, Jr., AMM3c, USNR, Air Gunner, Mech., Atlanta, Tex.

Crew No. 4 "FLYING CIRCUS"

Lt. Cmdr. John F. Bundy, USN, Pilot, PPC, Flight Officer, Sioux City, Iowa
Lt. (jg) Robert E. Malmfeldt, USN, PPC, Co-Pilot, New Buffalo, Mich.
Ens. James J. Hellesen, USNR, Navigator, Chicago, Ill.
Thomas H. Combs, AMM1c, USNR, Plane Captain, Air Gunner, Hazard, Ky.
Bennie C. Rubino, ACOM, USN, Air Gunner, Ordnance Chief, Ellwood City, Penna.
Donald P. Schoener, AOM2c, USN, Air Gunner, Ordnanceman, Rapid City, S. D.
Frank L. Smith, Jr., ARM2c, USNR, Radioman, Air Gunner, Raymond, Miss.
John J. Griffiths, ARM2c, USNR, Radioman, Air Gunner, Johnson City, N. Y.
George R. Bowly, AMM2c, USNR, Air Gunner, Mech., Menands, N. Y.
Ernest Marschke, Jr., AMM3c, USNR, Air Gunner, Mech., Milwaukee, Wis.
Alexander Stasinos, AMM3c, USNR, Air Gunner, Mech., Chicago, Ill.

Crew No. 5 "NO FOOLIN'"

Lt. Clifton B. Davis, USNR, Pilot, PPC, Alexandria, Va.
Lt. (jg) Harry B. Shepherd, USNR, Co-Pilot, Hanover, Mass.
Ens. Robert L. Bennett, USNR, Navigator, Manton, Mich.
Orval Stewart, AMM1c, USNR, Plane Captain, Air Gunner, Huron, S. D.
Warren A. Williams, AOM1c, USNR, Air Gunner, Ordnanceman, Saugus, Mass.
Bill E. Mincey, AMM2c, USN, Air Gunner, Mech., Kannapilis, N. C.
Wilbur V. Agee, ARM2c, USNR, Radioman, Air Gunner, Whittier, Calif.
Donald L. DeBruine, ARM2c, USNR, Radioman, Air Gunner, Sheboygan, Wis.
George A. Martin, AOM2c, USNR, Air Gunner, Ordnanceman, Waukesha, Wis.
William R. Smith, AOM3c, USNR, Air Gunner, Ordnanceman, Decatur, Ill.
Francis M. Walker, AMM3c, USNR, Air Gunner, Mech., Birmingham, Ala.

Crew No. 6 "CLIMBABOARD"

Lt. William B. Bridgeman, USNR, Pilot, PPC, Pasadena, Calif.
Ens. Robert P. Lamont, USNR, Co-Pilot, Washington Depot, Conn.
Ens. Herbert H. Rittmiller, USNR, Navigator, Torrance, Calif.
Edwin H. Watts, AMM1c, USN, Plane Captain, Air Gunner, Lincoln, Calif.
Peter T. Fay, ARM2c, USNR, Radioman, Air Gunner, Atlanta, Ill.
Walter J. McDonald, AMM2c, USNR, Air Gunner, Mech., Bell Gardens, Calif.
Howard E. Bensing, ARM2c, USNR, Radioman, Air Gunner, Louisville, Ky.
Robert A. Tovey, AOM2c, USNR, Air Gunner, Ordnanceman, Chicago, Ill.
George E. Murphy, AOM2c, USNR, Air Gunner, Ordnanceman, Milwaukee, Wis.
Warren G. Griffin, AMM3c, USNR, Air Gunner, Mech., Marietta, Okla.
Robert W. Carey, ARM3c, USNR, Radioman, Air Gunner, Atlanta, Ga.

Crew No. 7 "THE STORK"

Lt. Cmdr. George L. Hicks, USNR, Pilot, PPC, Executive Officer, Three Rivers, Calif.
Lt. William A. Warren, USNR, Co-Pilot, PPC, St. Paul, Minn.
Ens. Dicky Wieland, USNR, Navigator, Dearborn, Mich.
Joseph L. Oppenheimer, ACOM, USN, Leading Chief, no address
Charles A. Murphy, ACMM, USN, Plane Captain, Air Gunner, Watertown, Tenn.
Roger W. Clemons, AMM1c, USN, Plane Captain, Air Gunner, Columbus, Ohio
Sidney R. Metcalf, Chief ACOM, USN, Air Gunner, Ordnanceman, Framingham, Mass.
David H. Carlson, AOM1c, USN, Air Gunner, Ordnance, Everett, Wash.
William B. Idleman, AOM2c, USN, Air Gunner, Ordnanceman, Burbank, Calif.
Donald V. Conry, ARM2c, USNR, Radioman, Air Gunner, Fond Du Lac, Wis.
Bill N. Harris, ARM2c, USNR, Radioman, Air Gunner, Houston, Tex.
William A. McNeil, AMM2c, USNR, Air Gunner, Mech., Springfield, Ohio
Chester E. Rosell, AMM3c, USNR, Air Gunner, Mech., Leonardville, Kan.

Crew No. 8

Lt. John D. Keeling, USNR, Pilot, PPC, Scott City, Kan.
Ens. Alfred T. Peterson, USNR, Co-Pilot, Litchville, N. D.
Ens. Willard C. Hollingsworth, USNR, Navigator, Gary, Ind.
Donald R. Lursman, AMM1c, USNR, Plane Captain, Air Gunner, Walla Walla, Wash.
William F. Krier, ARM1c, USNR, Radioman, Air Gunner, Philadelphia, Pa.
Frank R. Kramer, AOM1c, USNR, Air Gunner, Ordnanceman, Bingen, Wash.
Dennis E. Bornholtz, AOM1c, USNR, Air Gunner, Ordnanceman, Sioux City, Iowa

Joseph W. Sarno, AMM1c, USNR, Air Gunner, Mech., Amityville, N. Y.
Robert D. West, ARM2c, USNR, Radioman, Air Gunner, Olathe, Col.
Donald M. Maloney, AMM2c, USN, Air Gunner, Mech., Ferndale, Mich.
Frank S. Graves, AOM2c, USNR, Air Gunner, Ordnanceman, Pekin, Ind.

Crew No. 9 "OUR BABY"

Lt. George A. Mellard, USNR, Pilot, PPC, Russell, Kan.
Lt. (jg) John Dooley, USNR, Co-Pilot, Boston, Mass.
Ens. Bruce D. DeJager, USNR, Navigator, Rochester, N. Y.
Andrew Halasz, AMM1c, USNR, Plane Captain, Air Gunner, Dorchester, Mass.
Thomas C. Lee, AOM1c, USNR, Air Gunner, Ordnanceman, Louisville, Ky.
Jack H. Renfro, AMM2c, USNR, Air Gunner, Mech., Reno, Nev.
Claude L. McWhorter, AOM2c, USNR Air Gunner, Ordnanceman, Danville, Ill.
Robert L. Dumais, ARM2c, USNR, Radioman, Air Gunner, Brunswick, Me.
John P. Hruska, ARM2c, USNR, Radioman, Air Gunner, Nahma, Mich.
Evans R. Lally, Jr., AMM3c, USNR, Air Gunner, Mech., San Francisco, Calif.
Frank P. Guarini, AMM3c, USNR, Air Gunner, Mech., Frankfort, Ill.

Crew No. 10 "COME GET IT"

Lt. Cmdr. William Janeshek, USN, Pilot, PPC, Port Washington, Wis.
Lt. (jg) Thomas C. Peebles, USNR, Co-Pilot, Newton Centre, Mass.
Ens. Asa L. Branstetter, USNR, Navigator, Curryville, Mo.
Albert C. Peterson, AMM1c, USN, Plane Captain, Air Gunner, Ft. Pierce, Fla.
Rolvin P. Allen, AMM2c, USNR, Air Gunner, Mech., Cranston, R. I.
Seymour Simon, ARM2c, USNR, Radioman, Air Gunner, Golden, Colo.
Walter G. Holmes, ARM2c, USNR, Radioman, Air Gunner, Linesville, Pa.
Joseph A. Yates, III, AOM2c, USNR, Air Gunner, Ordnanceman, Miami, Fla.
J. C. York, AOM2c, USN, Air Gunner, Ordnanceman, Saylor, Mo.
Jack T. Biggers, AOM3c, USNR, Air Gunner, Ordnanceman, Inglewood, Calif.
Joseph B. Rogers, AOM3c, USN, Air Gunner, Ordnanceman, Dallas, Tex.

Crew No. 11 "HELLDORADO"

Lt. Joseph I. Jadin, USNR, Pilot, PPC, Gladstone, Mich.
Lt. (jg) Charles W. Tischoff, USNR, Co-Pilot, Gloucester, Mass.
Ens. William F. Morgan, USNR, Navigator, Hollidaysburg, Penna.
Jack H. Cosby, AMM1c, USN, Plane Captain, Air Gunner, Willard, Mo.
Julius H. Loeser, AOM1c, USN, Air Gunner, Ordnanceman, Houston, Tex.
Richard Malkosian, AOM1c, USNR, Air Gunner, Ordnanceman, Westboro, Mass.
Emory L. Peterman, ARM2c, USNR, Radioman, Air Gunner, Cedar Falls, Iowa
Francis W. Farrell, AMM2c, USN, Air Gunner, Mech., New York, N. Y.
Amos Mellinger, ARM2c, USNR, Radioman, Air Gunner, Kalona, Iowa

Joseph M. McKernan, AMM3c, USNR, Air Gunner, Mech., Chicago, Ill.
Cecil M. Lee, AMM3c, USNR, Air Gunner, Mech., Calhoun City, Miss.

Crew No. 12

Lt. Harold E. Belew, USNR, Pilot, PPC, Fresno, Calif.
Lt. (jg) Robert W. Conkey, USNR, Co-Pilot, Pawtucket, R. I.
Ens. Alexander J. Bacon, USNR, Navigator, Detroit, Mich.
John F. Haznar, AMM1c, USNR, Plane Captain, Air Gunner, Thompsonville, Conn.
James L. Chase, ACRM, USN, Chief Radioman, Air Gunner, Miami, Fla.
Walter D. Morris, AOM1c, USN, Air Gunner, Ordnanceman, Eaton, N. Y.
Richard L. Duncan, ARM2c, USNR, Radioman, Air Gunner, Bedford, Ind.
Robert E. Spomer, AMM2c, USN, Air Gunner, Mech., Oakland, Calif.
Claude E. Watkins, AOM2c, USNR, Air Gunner, Ordnanceman, Wenatchee, Wash.
Arthur R. Sherren, AMM3c, USNR, Air Gunner, Mech., Waterloo, Iowa
Charles F. Kidwell, AMM3c, USNR, Air Gunner, Mech., Venice, Calif.

Crew No. 14

Lt. Leo E. Kennedy, USNR, Pilot, PPC, Ethlyn, Mo.
Ens. Marcus F. Poston, USN, Co-Pilot, El Paso, Tex.
Ens. Richard J. Potter, USNR, Navigator, Center Conway, N. H.
William E. White, AMM1c, USNR, Plane Captain, Air Gunner, Danville, Ga.
Troy A. McClure, AOM1c, USNR, Air Gunner, Ordnanceman, Cloverdale, Ala.
William L. Willocks, Jr., AMM2c, USNR, Air Gunner, Mech., Schenectady, N. Y.
Joseph B. Edwards, ARM2c, USNR, Radioman, Air Gunner, Des Moines, Iowa
Garnet R. Brown, AOM2c, USNR, Air Gunner, Ordnanceman, Jet, Okla.
John O. Oates, AMM3c, USNR, Air Gunner, Mech., Coronada, Calif.
Jacob E. Bryant, ARM3c, USNR, Radioman, Air Gunner, Newellton, La.
Harlyn G. Bakko, AMM3c, USNR, Air Gunner, Mech., Billings, Mont.

Crew No. 15 "URGE ME"

Lt. Oliver S. Glenn, USNR, Pilot, PPC, New Orleans, La.
Lt. (jg) John P. McKenna, USNR, Co-Pilot, Mt. Vernon, N. Y.
Ens. Howard L. Flohra, USNR, Navigator, Creston, Iowa
Richard H. Westmoreland, AMM2c, USNR, Plane Captain, Air Gunner, Wichita Falls, Tex.
Robert M. White, ARM2c, USNR, Radioman, Air Gunner, Springfield, Mass.
Clarence O. Gibson, AOM2c, USNR, Air Gunner, Ordnanceman, Lawrence, Ind.
James Grieve, AOM2c, USNR, Air Gunner, Ordnanceman, Chicago, Ill.
Joseph A. Mirr, AOM2c, USNR, Air Gunner, Ordnance, Milwaukee, Wis.

John A. Ryll, ARM2c, USNR, Radioman, Air Gunner, Mech., North Plymouth, Mass.
Lawrence L. Morino, AMM2c, USN, Air Gunner, Mech., Los Angeles, Calif.

Crew No. 16 "CONSOLIDATED'S MISTAKE"

Lt. Joseph H. Jobe, USNR, Pilot, PPC, Goldendale, Wash.
Lt. (jg) Raymond T. Miller, USNR, Co-Pilot, Seymour, Tex.
Ens. Oscar Juelke, USNR, Navigator, Oaks, N. D.
Henson Guernsey, AMM2c, USNR, Plane Captain, Air Gunner, Bivins, Tex.
James F. Curtis, AMM1c, USNR, Air Gunner, Mech., Indianapolis, Ind.
Wesley L. Thompson, ARM2c, USNR, Radioman, Air Gunner, Columbus, Kan.
Armond L. Klotz, ARM2c, USNR, Radioman, Air Gunner, Vinita, Okla.
Louis F. Paulukonis, AOM2c, USNR, Air Gunner, Ordnanceman, Far Rockaway, N. Y.
Robert A. Simms, AOM2c, USNR, Air Gunner, Ordnanceman, Buda, Ill.
Harold J. Carter, AMM3c, USNR, Air Gunner, Mech., Okmulgee, Okla.
Gustav Greve, AMM3c, USNR, Air Gunner, Mech., Annapolis, Md.

Crew No. 17 "AVAILABLE JONES"

Lt. Oden E. Sheppard, USNR, Pilot, PPC, Bozeman, Mont.
Ens. Clifford I. Nettleton, USN, Co-Pilot, PPC, Independence, Kan.
Ens. Melvin M. Breitsprecher, USNR, Navigator, Waterloo, Iowa
Edward M. Danilczuk, AMM1c, USN, Plane Captain, Air Gunner, Bridgeport, Conn.
Robert C. Thompson, ARM2c, USNR, Radioman, Air Gunner, Tulsa, Okla.
Robert E. Bartlett, AMM2c, USNR, Air Gunner, Mech., Corpus Christi, Tex.
Carl R. Snyder, AOM2c, USNR, Air Gunner, Ordnanceman, Carthage, Mo.
Howard L. Haynes, AOM2c, USNR, Air Gunner, Ordnanceman, Watertown, Tenn.
Thomas M. Leamons, ARM2c, USNR, Radioman, Air Gunner, Huntington Park, Calif.
Howard R. Anthony, AOM3c, USNR, Air Gunner, Ordnanceman, Ft. Worth, Tex.
Elmer C. Webb, AMM3c, USNR, Air Gunner, Mech., Purdow, Tex.

Crew No. 18 "SUGAR QUEEN"

Lt. Thomas R. Wheaton, USNR, Pilot, PPC, San Diego, Calif.
Lt. (jg) Ernest B. Anderson, USNR, Co-Pilot, New Haven, Conn.
Ens. Frederick J. Menninger, Jr., USNR, Navigator, Berkeley, Mich.
Raymond E. McPeak, AMM1c, USN, Plane Captain, Air Gunner, Mt. Airy, N. C.
Louie J. Thomas, ACOM, USN, Air Gunner, Ordnance Chief, Alabama City, Ala.
Gordon V. Johnson, AOM2c, USNR, Air Gunner, Ordnanceman, Highland Park, Ill.

J. L. Howard, ARM2c, USN, Radioman, Air Gunner, Los Angeles, Calif.
Jack R. Musburger, AOM2c, USNR, Air Gunner, Ordnanceman, Bemidji, Minn.
Joseph Beregi, AMM2c, USNR, Air Gunner, Mech., Bridgeport, Conn.
Vernon A. Elchert, ARM2c, USNR, Radioman, Air Gunner, Fostoria, Ohio.
Ernest S. Hayes, AOM3c, USNR, Air Gunner, Ordnanceman, El Cajon, Calif.

Crew No. 20 "EARTHBOUND"

Lt. Richard C. Crouch, USNR, Personnel Officer
Lt. Theodore M. Steele, USNR, Air Combat Intelligence and Public Relations Officer
Lt. William W. Robinson, USNR, Operations Officer
Lt. Lyman W. Ballinger, USNR, Engineering Officer
Lt. Max B. Payne, USNR, Gunnery Officer
Chester S. Jones, CY, USNR, Yeoman
John S. Sladich, Y1c, USNR, ACI Yeoman

OFFICERS DETACHED DURING CRUISE

Ens. Walter L. Eckalbar, Jr., USNR
Ens. Harold T. Fox, USNR
Ens. Harry Bigham, USNR
Ens. Eugene D. Hadlock, USNR
Ens. Ira Smith, Jr., USNR
Lt. Jackson L. Grayson, USN

Ens. Robert R. Krenrick, USNR
Lt. Harold M. McCarthy, USNR
Lt. Edward W. Lloyd, USNR
Lt. (jg) Norman L. Burton, USNR
Ens. William T. Wiley, USNR

MEN DETACHED DURING CRUISE

E. O. Ray, S2c, USNR
R. R. Summers, AOM2c, USNR
F. T. Dempsey, ARM2c, USNR
Edward C. Mazewski, AMM1c, USN

Alfred W. Dibb, AMM2c, USNR
Howard C. Spilker, AOM2c, USNR
Philip F. Lininger, AOM2c, USNR
Lloyd B. Bowen, AMM2c, USNR

MISSING IN ACTION
13 January 1944

Lt. Samuel E. Coleman, USNR, Pilot, PPC
Lt. (jg) Amos L. Shreiner, USNR, Co-Pilot
Ens. Leslie E. Fontaine, USNR, Navigator
Louie E. Sandidge, Jr., AMM1c, USNR, Plane Captain, Air Gunner
Sterling F. Brown, AMM2c, USNR, Air Gunner, Mech.
Harry F. Donovan, ARM2c, USNR, Radioman, Air Gunner
Daniel J. Dujak, ARM2c, USNR, Radioman, Air Gunner
James T. Heasley, AOM2c, USNR, Air Gunner, Ordnanceman
Truman Steele, AOM2c, USNR, Air Gunner, Ordnanceman
Leo C. Petrick, S1c, USNR, Air Gunner
John Tusha, S1c, USNR, Air Gunner

212 I TOOK THE SKY ROAD

13 February 1944

Lt. (jg) John H. Herron, USNR, Pilot, PPC
Lt. (jg) Charles M. Henderson, Jr., USNR, Co-Pilot
Ens. Nelson T. O'Bryan, USNR, Navigator
Robert J. Bennington, AMM1c, USNR, Plane Captain, Air Gunner
Benjamin W. Anderson, AOM(T)1c, USNR, Air Gunner, Ordnanceman
Raymond W. Devlin, ARM2c, USNR, Radioman, Air Gunner
Andrew W. Piscopo, ARM2c, USNR, Radioman, Air Gunner
Clarence H. Ziehlke, AMM3c, USN, Air Gunner, Mech.
Paul J. Graham, AOM3c, USNR, Air Gunner, Ordnanceman
Richard C. Vancitters, AOM3c, USNR, Air Gunner, Ordnanceman
Aern R. Durgin, S2c, USNR, Air Gunner

5 August 1944

Lt. Elmer H. Kasperson, USN, Pilot, PPC
Ens. Warren A. Hindenlang, USNR, Co-Pilot
Ens. Keith E. Ellis, USNR, Navigator
Joseph W. Komorowski, AMM1c, USNR, Plane Captain, Air Gunner
Warren B. Simon, AMM2c, USN, Air Gunner, Mech.
Richard D. Frye, ARM2c, USNR, Radioman, Air Gunner
Hugo L. Kluge, ARM2c, USNR, Radioman, Air Gunner
William F. Schneider, AOM2c, USNR, Air Gunner, Ordnanceman
Victor B. Jones, AOM2c, USN, Air Gunner, Ordnanceman
Allen K. Stinger, AOM3c, USNR, Air Gunner, Ordnanceman
Bobby W. Fickling, AOM3c, USNR, Air Gunner, Ordnanceman

KILLED IN ACTION
5 June 1944

Hale D. Fisher, AMM1c, USNR, Plane Captain, Air Gunner

The XPBS-1—Pacific trail-blazer.

The *Thunder Mug's* last crew. Front: Lawrence Johnson, Thomas Delahoussaye, Wayne Young, Jack Simmen. Rear: Edwin Dorris, Ensign Robert K. Schaffer, Commander Miller, Robert Gariel, Bernard Jaskiewicz.

The *Thunder Mug*. Sixty-six Jap ships felt her sting.

Vice Admiral John H. Hoover, USN, presents to Commander Miller his first and second Air Medals.

a surprise low-level visit to Kwajalein, before the occupation of that island by our troops, the Liberators of VB-109 left these and other Jap ships furiously blazing.

A Japanese coastal vessel, caught by Lieutenant Seabrook near Toangi, burned briskly before going down.

Lieutenant Kasperson found this enemy freighter and escort on the high seas. Scratch one freighter.

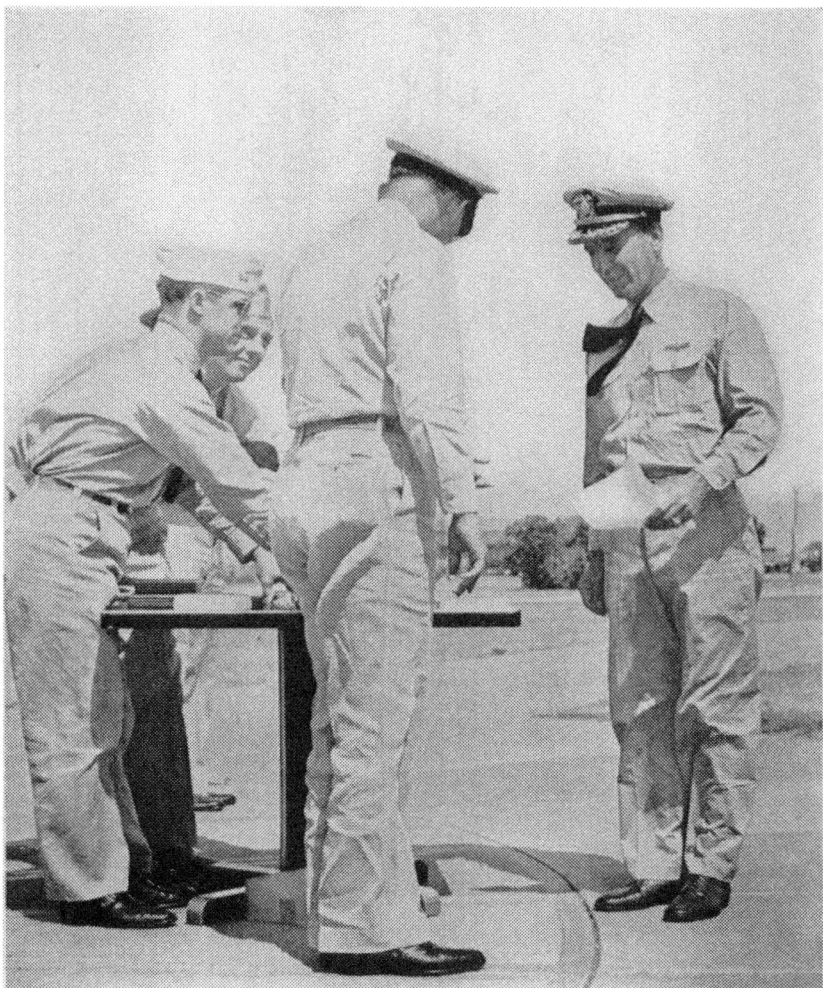

Rear Admiral John Dale Price, USN, Commander Fleet Air Wing Two (back to camera), is reaching for the Gold Star which he presented to Commander Miller in lieu of a third Navy Distinguished Flying Cross.

The squadron pays its last respects to Hale D. Fisher, killed in action.

Spectacular death of a Jap Mavis. Lieutenant Keeling's crew made the kill after the Jap was out-maneuvered in an exciting 19-minute chase through the clouds.

The discovery of this air strip marked the beginning of Commander Miller's "one-man-war" on Puluwat Island.

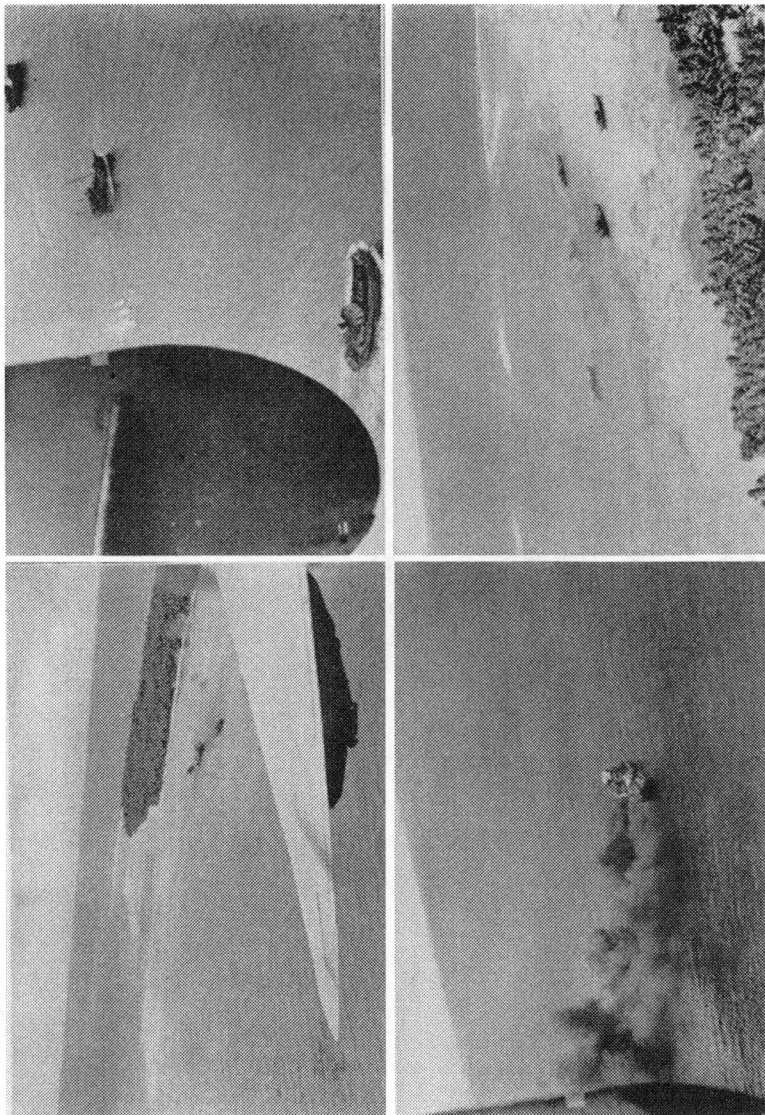

The Birth of a Legend. On a "routine search" of the Hall Islands, Commander Miller discovered five enemy ships at Ruo and three more at near-by Murilo. He sank four of the five and two of the three, for a total of six. The other two were "damaged."

The Japs at Nomwin lose another of their important radio- and weather-stations.

Ushi Point airfield, Tinian. The devastation was complete.

The tower at Puluwat. Commander Miller had reason to remember it.

When a Jap shell burst above the plane at Puluwat, the top-turret guns were depressed and fired into the cockpit, occupied by Commander Miller and Lieutenant Bill Bridgeman.

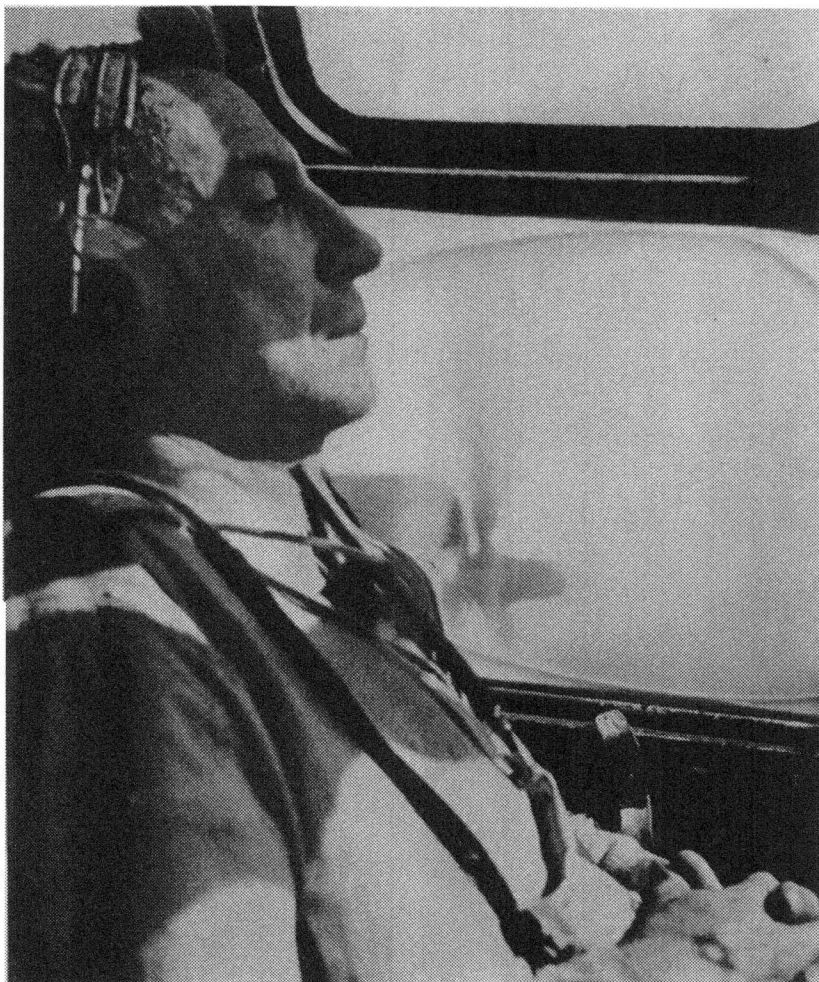

A dramatic interpretation of the phrase "In the Highest Traditions of the Naval Service." Wounded over Puluwat, Bus Miller remained at the controls of a badly shot-up Liberator for seven hours and brought plane and crew safely back to base. The picture was snapped through the cockpit window, a moment after the plane landed. Wounds are plainly visible on hand and temple. "It was a long way home," the commander says.

Journey's End. On her last trip to Ponape, the *Thunder Mug* returned crippled to Eniwetok, landed without brakes, and rolled into the sea. No one was hurt.